SpringerBriefs in Environmental Science

SpringerBriefs in Environmental Science present concise summaries of cutting-edge research and practical applications across a wide spectrum of environmental fields, with fast turnaround time to publication. Featuring compact volumes of 50 to 125 pages, the series covers a range of content from professional to academic. Monographs of new material are considered for the SpringerBriefs in Environmental Science series.

Typical topics might include: a timely report of state-of-the-art analytical techniques, a bridge between new research results, as published in journal articles and a contextual literature review, a snapshot of a hot or emerging topic, an in-depth case study or technical example, a presentation of core concepts that students must understand in order to make independent contributions, best practices or protocols to be followed, a series of short case studies/debates highlighting a specific angle.

SpringerBriefs in Environmental Science allow authors to present their ideas and readers to absorb them with minimal time investment. Both solicited and unsolicited manuscripts are considered for publication.

More information about this series at http://www.springer.com/series/8868

Kullapa Soratana • Amy E. Landis
Fu Jing • Hidetsugu Suto

Supply Chain Management of Tourism Towards Sustainability

 Springer

Kullapa Soratana
Faculty of Logistics and Digital Supply
Chain
Naresuan University
Phitsanulok, Thailand

Amy E. Landis
Colorado School of Mines
Golden, CO, USA

Fu Jing
Chengdu University
Chengdu, China

Hidetsugu Suto
Muroran Institute of Technology
Muroran, Japan

ISSN 2191-5547   ISSN 2191-5555   (electronic)
SpringerBriefs in Environmental Science
ISBN 978-3-030-58224-1   ISBN 978-3-030-58225-8   (eBook)
https://doi.org/10.1007/978-3-030-58225-8

This Springer imprint is published by the registered company Springer Nature Switzerland AG
The registered company address is: Gewerbestrasse 11, 6330 Cham, Switzerland

Preface

Supply Chain Management of Tourism Towards Sustainability is a collection of approaches to drive the supply chain of the tourism industry towards sustainability. The purpose of this book is to offer guidance for students, teachers, researchers, school management, and/or corporations who are interested in applying sustainability into tourism-related activities. Students, teachers, researchers, and/or corporations can adopt the contents and the proposed integrated framework of supply chain management (SCM) and life cycle approaches (LCA) in this book as initiatives to further develop their studies or apply and implement them in practice.

The approaches discussed in this book are SCM, LCA, knowledge supply chain (KSC) model, benefit of inconvenience (BI), media biotope (MB), and the proposed integrated framework of SCM and LCA. Each approach can be adopted to improve tourism activities, tourists' experiences, and regional development from a sustainability perspective. LCA (Chap. 2) can be used to quantify environmental impacts and cost over the entire life cycle of products or services. The LCA results can be used in process improvement and in the decision-making process. Process improvement and knowledge and best practices among stakeholders in the tourism sector can be shared through the KSC model (Chap. 3). Another important element in achieving sustainable development of the tourism industry is tourists' experiences, which can be promoted by using BI of MB (Chap. 4). A combined LCA-SCM framework delivers a new insight into the sustainable development of the tourism industry. How resources and/or operations used in different sectors across the tourism industry or among suppliers can be shared to reduce the costs of resources and operation by using the proposed LCA-SCM framework is discussed in Chap. 5.

This book also discusses tradeoff issues among the three pillars of sustainability—environment, society, and economic—from implementing sustainable tourism throughout its supply chain and provides good practices on how to keep the three pillars in balance. The issues in the sustainable development of tourism and how each approach can be incorporated in the tourism industry are presented in the five chapters as follows:

Chapter 1 aims at introducing why sustainable development in tourism is in the spotlight and how the problems should be managed. Information on how sustainable development in the tourism sector has become a major global issue is provided. The effects of climate change, one of the key driving forces, are discussed. The approaches, which are SCM and LCA, to ease the unsustainable development situation in the tourism sector are introduced in this chapter.

Chapter 2 consists of four main sections. The first two sections are: an introduction of life cycle approaches to inform the readers on the major benefits of life cycle approaches and steps in conducting LCA. The third section reveals examples of constructing LCA for tourism activities such as accommodation, transportation, restaurants, tour operators, and waste disposal facilities. The section aims to provide guidance on how to conduct an LCA. The fourth section discusses how LCA can be applied to each activity over the entire life cycle of a tourism service, i.e. restaurants.

Chapter 3 is on KSC as an appropriate process to manage knowledge flow in designing an effective ICT-empowered tourism knowledge supply chain. KSC is an integration of knowledge management and supply chain management. KSC is an important process in integrating educational provisions with government and industry needs through preparing visions, minds, and skills to support regional development, which is sustainability. The tourism sector can adopt and use KSC as a platform for knowledge provision and as a domain for knowledge sharing.

Chapter 4 discusses the design criteria of BI and communication model of MB. Both BI and MB potentially promote the social sustainability aspect of tourism. BI can be used to design tourism activities that enhance creativity and raise awareness on social sustainability. For example, some tourists prefer creating local crafts despite they can purchase similar products from souvenir shops. MB system is referred to a structure of communication media patterns in communities. The BI of MB, created by MB mediums or biotope-oriented mediums, can be used to maintain the originality and sustainability of the communities or tourist destinations.

Chapter 5 proposed an integrated framework of LCA and SCM for tourism. The proposed LCA-SCM framework for tourism is discussed. The opportunities and caveats of LCA and SCM for tourism enterprises, such as accommodation, restaurant, and transportation, are specified.

Phitsanulok, Thailand   Kullapa Soratana
Golden, CO, USA   Amy E. Landis
Chengdu, China   Fu Jing
Muroran, Japan   Hidetsugu Suto

Contents

1 Sustainable Development of Tourism 1
  1.1 The Importance of Sustainable Development in Tourism
      Sector 1
  1.2 Unsustainable Development in the Tourism Sector 3
      1.2.1 A Top-Down Policy Based-Tourism 4
      1.2.2 Greenwashed Ecotourism 5
  1.3 Overview of Sustainability-Driven Approaches 6
      1.3.1 Supply Chain Management Approach 7
      1.3.2 Life Cycle Approach 8
  1.4 Tourism Enterprises During a Crisis to Avoid Unsustainable
      Tourism Activities 9
      1.4.1 Pre-crisis 10
      1.4.2 During the Crisis 10
      1.4.3 Post-crisis 10
  References 11

**2 The Role of Life Cycle Approaches in Sustainable Development
  of Tourism** 13
  2.1 An Introduction to Life Cycle Approaches 13
  2.2 Life Cycle Assessment and Its Implication 14
      2.2.1 Goal and Scope Definition 15
      2.2.2 Life Cycle Inventory 16
      2.2.3 Life Cycle Impact Assessment 17
      2.2.4 Interpretation and Improvement Analysis 18
  2.3 Tourism Activities Over Their Life Cycle 19
      2.3.1 Accommodations 20
      2.3.2 Transportation 21
      2.3.3 Restaurants 22
      2.3.4 Tour Operators 23
      2.3.5 Waste Disposal Facilities 25

2.4   Good Practices in Using LCA to Promote Sustainable Development
       Tourism 25
References.. 27

**3   A Knowledge Supply Chain for the ICT-Enhanced Tourism
      Education** 31
3.1   ICT Usage in the Tourism Sector. 31
3.2   KSC: The First Time to Integrate KM and SCM in the Context
       of Tourism and Its Education................................... 32
       3.2.1   KE, SCM, and Associated Tools 32
       3.2.2   KM and Its Tools 32
       3.2.3   A Knowledge Supply Chain (KSC): Integrating
                SCM and KM... 33
3.3   Closing the Gap Between Knowledge Supply and Demand
       in e-Tourism Curriculum Design 33
       3.3.1   Data Collection and Structuring. 33
       3.3.2   Benchmarking 37
       3.3.3   Understanding the "As-Is" Situation 38
       3.3.4   Simulating the Desired "to-Be" Situation 38
3.4   Conclusions 40
References.. 41

**4   Improvement of Tourists' Experience to Promote Sustainable
      Tourism** 43
4.1   Benefit of Inconvenience and Sustainable Tourism Design. 43
       4.1.1   Benefit of Inconvenience 43
       4.1.2   Media Biotope 45
       4.1.3   Systems for Sustainable Tourism Based
                on BI and MB.. 48
       4.1.4   Conclusions 53
References.. 53

**5   Life Cycle Assessment and Supply Chain Management
      for Tourism Enterprises** 55
5.1   Combination of LCA and SCM Aspects for Sustainable
       Development 55
5.2   The Opportunities and Limitations of LCA-SCM for Tourism
       Enterprises 57
       5.2.1   Accommodation....................................... 57
       5.2.2   Restaurant 60
       5.2.3   Transportation 61
References.. 63

Abbreviations

3Rs	Reduce, reuse and recycle
AIRS	Air quality planning and standard
B2B	Business-to-business
BEES	Building for environmental and economic sustainability
BI	Benefit of inconvenience
BREEAM	Building research establishment environmental assessment method
CAMT	College of Arts, Media and Technology
CFC-11	Chlorofluorocarbon-11
CMU	Chiang Mai University
CoP	Communities of practices
CSR	Corporate social responsibility
EIA	Energy Information Administration
EPD	Environmental product declaration
EVs	Electric vehicles
GDP	Gross domestic production
GHGs	Greenhouse gas emissions
GMS	Greater Mekong sub-region
GSTC	Global Sustainable Tourism Council
HE	Higher education
ICTs	Information and communication technologies
IPCC	Intergovernmental panel on climate change
ISO	International Organization for Standardization
IT	Information technology
KADS	Knowledge analysis and data structuring
KE	Knowledge engineering
KM	Knowledge management
KMS	Knowledge management system
KSC	Knowledge supply chain
LCA	Life cycle assessment
LCC	Life cycle costing
LCI	Life cycle inventory analysis

LCIA	Life cycle impact assessment
LCSA	Life cycle sustainability assessment
LEED	Leadership in energy and environmental design
MB	Media biotope
SCM	Supply chain management
SCOR	Supply chain operations reference
SDGs	Sustainable development goals
S-LCA	Social-life cycle assessment
SMEs	Small and medium sized enterprises
TRACI	Tool for reduction and assessment of chemical and other environmental impacts
TRI	Toxic release inventory
TRL	Tourism revenue leakage
UNESCO	United Nations Educational, Scientific and Cultural Organization
UNWTO	United Nations World Tourism Organization

Chapter 1
Sustainable Development of Tourism

1.1 The Importance of Sustainable Development in Tourism Sector

Climate change is a direct threat to tourism destinations. Climate change can be defined as: *"any change in climate over time, whether due to natural variability or as a result of human activity"* in IPCC usage, or *"a change of climate that is attributed directly or indirectly to human activity that alters the composition of the global atmosphere and that is in addition to natural climate variability observed over comparable time periods"* in the Framework Convention on Climate Change usage [1]. Evidences of climate change impacting tourism destinations are seen in coral bleaching in at least 10 popular diving sites of Thailand, glacier shrinking in, e.g., Greenland, Norway, and Iceland, and massive forest fires in Australia and Amazon rainforest in Brazil, Bolivia, Peru, and Paraguay [2–4]. A more comprehensive review on impacts of climate change on tourism destinations can be found in, for instance, the studies by Scott, D. et al., Santarius, T. et al., and Aall, C. et al. [5–7]. The impacts do not only deteriorate ecosystems and nature, but also affect the society and the economy, which are the three pillars of sustainability. In 2018, the tourism sector accounted for 10.4% of the world gross domestic product (GDP) and 319 million jobs or 10% of total employment. The direct GDP and direct employment accounted for 3.3% and 3.9%, respectively [8]. Thus, the reduction of number of tourists to a tourism destination due to climate change catastrophes negatively affects direct and indirect tourism businesses and employment.

The direction of tourism development is crucial. The tourism sector involves with a wide array of businesses and activities that form a services network. The network within a tourist destination consists of services such as accommodations, restaurants, transport of humans and goods, shops, tour operators, and travel agencies. The network also includes supportive businesses such as producers of food and beverage, equipment and furniture, and waste recycling and disposal facilities [9].

© The Author(s) 2021
K. Soratana et al., *Supply Chain Management of Tourism Towards Sustainability*, SpringerBriefs in Environmental Science,
https://doi.org/10.1007/978-3-030-58225-8_1

Although each tourism-related business and activity generates income and job opportunities, expecting to improve residents' quality of life, such business and activities consume natural resources and generate solid wastes, wastewater, and greenhouse gas emissions. Based on the existing operation of tourism industry through 2050, the sector would consume 154% more of energy and consume 152% more of water, and would generate 131% more of greenhouse gas emissions and 251% more of solid wastes [10]. The total resource consumption, e.g. energy, freshwater, and land use, of the tourism sector for the period 1900–2050 is reported in the study by Gössling, S. and P. Peeters [11]. Overconsumption of natural resource and generation of too much waste could result in eternal environment depreciation of tourism destinations, which are fundamental resources of tourism industry. Therefore, the development of tourism should be directed towards sustainability.

The term sustainable development of tourism was defined by United Nations World Tourism Organization (UNWTO) [12] as:

> Tourism that takes full account of its current and future economic, social and environmental impacts, addressing the needs of visitors, the industry, the environment and host communities.

Based on the definition, all three pillars of sustainability—economy, society, and environment—must be taken into consideration and kept in balance for any operation in the tourism sector. The goal of sustainable development is to enhance economic stability of businesses while maintaining a well-functioning society and a healthy environment [12, 13]. A concurrent consideration of two pillars is an initiative to ultimately achieve sustainability. The balancing between society and economic leads to *equitable* condition, where resources are equally and fairly shared among people in the community. The balancing between economic and environment leads to a *viable* condition, where operations are pursued to meet both economic growth and environmental protection. The balancing between society and environment leads to *bearable* condition, where society works with awareness on environmental impacts and well-being of the community. In practice, keeping all three pillars of sustainability in balance is not simpler. The community that values more on the environment would set certain limitations of wastes and emissions to protect environment. The limitations may cause some difficulties for economic growth.

Sustainable development of tourism is envisaged as a global issue. The importance of sustainable development of tourism at global level is revealed in three Sustainable Development Goals (SDGs). SDGs are the common goal of United Nations Members since 2015 to transform our world into a more sustainable world by 2030 [14]. These three SDGs are Goal 8—to promote sustainable tourism that creates jobs and promotes local culture and products via policies, Goal 12—to develop tools to monitor and assess both positive and negative impacts from tourism operation, and Goal 14—to promote sustainable use of marine resources to avoid climate change and to promote well-being of people, respectively. However, it should be noted that current fast tourism development may conflict with SDGs, e.g. Goal 6 (Clean water and sanitation), Goal 7 (Affordable and clean energy), Goal 13 (Climate action), Goal 14 (Life below water), and Goal 15 (Life on land).

To successfully drive tourism sector towards sustainability, after setting clear goals and enforcing policies [15], a constructive process must be established. The process could be established based on results from supply chain management (SCM) and life cycle (LC) approaches. SCM and LC approaches are prominent sustainable development approaches [9, 16–20]. Supply chain management focuses on planning and management of activities from upstream to downstream processes. Tourism supply chain can be categorized into *before the trip* and *during the trip*. Activities *before the trip* are, for example, tour operator, travel agency, and transportation between the tourist's home and the destination. Activities and services *during the trip* are, for example, sourcing and procurement, accommodation, catering, transportation, and tourist's activities. Another approach discusses in this book is LC approach. LC approaches consist of life cycle assessment (LCA), life cycle costing (LCC), and life cycle sustainability assessment (LCSA). Stages considered in LC approaches are from raw material acquisition to manufacturing, transportation, use, and end-of-life phase. LCA and LCC are applied to evaluate environmental impacts and costs over a life cycle of products or services, respectively [21]. Both LC approaches help identifying processes with major resource consumption and/or impact contribution and preventing rebound effects. LCSA is an approach to evaluate production and consumption impacts on all actors along the value chain [22]. A brief introduction on SCM and LCA is provided in Sect. 1.3 of this chapter.

1.2   Unsustainable Development in the Tourism Sector

In this section, unsustainable development in tourism sector is discussed on two levels: policy-making level and implementation level. Government plays a major role at the policy-making level, while several stakeholders, e.g. tour operator, lodging, and local community, operate at the implementation level. Both levels are directly related. Generally, stakeholders at the implementation level tend to follow the policy launched by the government. Most tourism development policies mainly aim to drive gross domestic product (GDP), an indicator of a country's economic growth. However, the policy with the focus solely on GDP does not assure a sustainable development of tourism [8]. The economy-driven tourism is mainly to spur a country's economy, while natural resources may be overexploited [23]. Public opinions about inadvertent environmental behavior play the role of enforcing action by the government, and, therefore, to avoid failure at implementation level, public opinions should be taken into consideration. In addition, a guideline on what should be conducted at implementation level is necessary. A top-down policy without social involvement prior to the policy-making process and sustainable development guideline may lead to unsustainable development of tourism.

1.2.1   A Top-Down Policy Based-Tourism

There are several cases of unsustainable development of tourism due to the lack of social involvement in top-down policy. One such case is Kiriwong Village in Nakhon Si Thammarat Province of Thailand [2]. Approximately 64% of the total population of the village are farmers, cultivating mangosteens and durians—tropical fruits, and another 10–15% of the total population are in tourism industry. The village received Thailand Tourism Awards and was recognized as a village of sustainable tourism. The government of Thailand promoted the village as tourism destination aiming to boost Thailand's GDP growth and resident's income. Despite several hundred tourists visiting the village each day brought by the promotion, tourism industry has overexploited water resources, generated solid wastes, and crowded the village. The environment and the whole community of the village are affected by water scarcity due to overtourism, while only a small number of the total population benefits from tourism.

Another continuing tourism development project in Thailand is the inno-life tourism-based community project. The inno-life tourism-based community project can be described as tourism activities that embrace innovations while preserving traditional way of life and promoting local products [24]. This project is, like many other policies, also initiated from a top-down management policy aiming to generate resident's income distribution. According to the project's notion, for a tourism destination to sustainably attract tourists, it needs to maintain its genuine value and condition. However, the project does not present or elevate the uniqueness of the community. Without any interesting uniqueness of the community to bring in and bring back visitors, the event, under the inno-life tourism-based project, organized by the government to promote the destination will only help resident's income distribution once, which is unsustainable. Therefore, varying features of tourism destinations should be exposed and have them imprinted in tourist's mind. Yet the community is willing to join the project with a lack of knowledge and know-how to develop their products or services. The project's budget should be invested on providing knowledge for a long-term development, rather than on building a check-in point for Instagram, Facebook, or other social media. Each small-scale construction of check-in point could cost up to 250,000 THB [25]. Unsustainable development from large-scale construction can be seen in Spain, one of the Mediterranean's mass tourism destinations. The constructions in Spain such as hotels, apartments, and concrete beachfront intended to attract tourists yield an adverse effect [26]. The place has shifted its tourists' orientation from nature to nightlife. Therefore, a bottom-up management policy approach coupled with tourism destination assessment on its management system and resource consumption should be implemented prior to further development.

1.2.2 Greenwashed Ecotourism

Ecotourism, one of the well-known types of tourism, in some cases, could cause unsustainable development. The International Ecotourism Society (TIES) defined ecotourism in 2015 as "responsible travel to natural areas that conserves the environment, sustains the well-being of the residents, and involves interpretation and education" [27]. According to TIES, ecotourism: is non-consumptive/non-extractive, creates an ecological conscience, and holds eco-centric values and ethics in relation to nature. Fennell notes that use of ecotourism originated in the 1960s, and since then definitions have varied greatly [28]. Most definitions have similar themes noted by TIES, including: having low environmental impact caused by visitors, providing revenue to local communities, promoting conservation, and incorporating an educational component for visitors. There are other forms of ecotourism, including [29]:

- Ethical tourism: focused on destinations where tourists learn about local ethical issues (e.g., social injustice and animal welfare).
- Geotourism: focused on enhancing the geographical characteristics of a place.
- Pro-poor tourism: focused on benefit the poor people in a destination.
- Responsible tourism: focused on maximizing benefits to local communities and their cultures and surrounding habitats.

Ecotourism enhances visitor awareness of environmental issues, and research shows that it may reinforce visitor's attitudes and behaviors toward environmental issues, result in post-trip lifestyle change, and increases their interest in further ecotourism experiences [16, 30, 31]. The local economic benefits, however, are not always in favor of the local community [32, 33]. In some cases, ecotourism offers new economic opportunities, but the economic benefits "may also reproduce preexisting patterns of stratification" and "may also engender processes of ideological resistance and reconfiguration that transform existing relationships of nationality, class, and gender" [34]. Researchers note that shallow, or greenwashed, ecotourism makes green claims in advertising with profit taking precedence over ecological and local community considerations [28]. "If not properly planned, managed, and monitored, ecotourism can be distorted for purely commercial purposes and even for promoting ecologically-damaging activities by large numbers of tourists in natural areas" [35]. An important distinction between ecotourism and sustainable tourism is that the former is an alternative, nature-based type of tourism focused on education of visitors, while sustainable tourism focuses on integrating sustainability into the entire tourism supply chain [35].

Unsustainable development issues of tourism require different management approaches. SCM can be adopted to resolve issues related to planning and management of tourism, while LCA can be adopted to resolve issues related to resource consumption and impact reduction of tourism activities. Some unsustainable development issues of tourism should be managed by SCM informed by LCA. Based on the example on Kiriwong Village and greenwashed ecotourism, overtourism and overexploitation of natural resource could be managed by SCM and LCA aimed at promoting sustainable development. SCM can assist in managing multiple suppliers in tourism's supply chain to develop an inter-organizational strategy [16]. LCA can assist in improving resource consumption efficiency and cost and impacts reduction of each tourism component. The six tourism service components, drawing tourists to tourism destinations, are attraction, accommodation, restaurant/café, products/souvenirs, transportation, and tourism activities, as illustrated in Fig. 1.1.

Each tourism service component has an important role in supporting tourism sector. They support a process in creating tourism logistics; physical flow (tourists and goods), financial flow, and information flow of tourism. For example, transportation service providers create physical flow through their buses and routes, while souvenirs shops create financial flow through their selling. A combination of the six components under sustainable development plan would create sustainable development in tourism sector. Conversely, sustainable development of only one of the six components could not drive tourism sector towards sustainability.

1.3   Overview of Sustainability-Driven Approaches

The major goal of sustainable development is to protect the environment under a profitable condition for economy and a bearable condition for society. Environmental protection as defined by the United Nations refers to "any activity to maintain or restore the quality of environmental media through preventing the emission of pollutants or reducing the presence of polluting substances in environmental media" [36]. According to the definition of environmental protection, environmental resources can still be consumed and utilized with consciousness to minimize and avoid environmental impacts.

Sustainable development should be driven based on supply chain and life cycle approaches. Both approaches take stakeholders related to tourism supply chain and over the life cycle of each service or product into consideration. The results could aid in policy decision-making level, rather than launching a policy relying solely on government or community's desire. However, obtaining reliable results from SCM and LCA requires stakeholders involvement and life cycle inventory development.

Fig. 1.1 Tourism service components drawing tourists to tourism destinations [by Kullapa Soratana]

1.3.1   Supply Chain Management Approach

SCM in tourism industry is a set of approaches utilized to efficiently manage and coordinate across multi-organizations in a network of tourism supply chain. The concept of tourism supply chain was introduced by UNWTO in 1975, focusing on the distribution networks and marketing activities in the tourism industry. Currently, researches have been expanded into seven key tourism SCM issues which are demand management, two-party relationship, supply management, inventory management, product development, tourism supply chain coordination, and information technology. Further information on each key tourism SCM issues can be read in a comprehensive review on tourism supply chain management by Zhang et al. [9].

Tourism products are complex. A tourism product can be achieved through a combination of different services. Therefore, tourism organizations need to consider beyond their market structure. The services in tourism supply chain are, e.g., reservation systems, transportation to and from destination, transportation around destinations, local transport and attraction providers, excursion, facilities, and suppliers to support local businesses, as illustrated in Fig. 1.2. All the services are inter-connected.

Tourism industry should be analyzed and managed from an integrated perspective to meet tourist satisfaction. There are four key components in tourism value chain, which are tourism supplier, tour operator, tour agent, and tourists. Tourism supplier can be categorized into Tier 1 and Tier 2 for different groups of supplies. Tier 1 suppliers are, e.g., accommodation, transportation, restaurant, and tourism activities. Tier 2 suppliers provide products and services to support Tier 1 supplier. Examples of Tier 2 suppliers are food and beverage manufacturers, furniture manufacturers, water and energy supplies, and waste recycling and disposal facilities. In addition, tourism industry can implement sustainable development by finding an equilibrium point between the consumption and the preservation of environmental

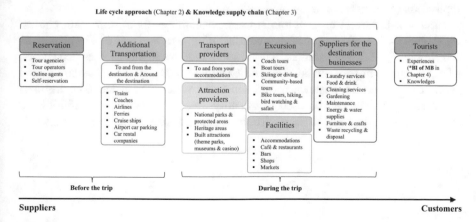

Fig. 1.2 Tourism supply chain. * Benefit of inconvenience of media biotope

resources (more details are provided in Chap. 2). Therefore, an adoption of SCM is appropriate to manage and connect suppliers through, i.e., knowledge sharing platform (more details are provided in Chap. 3) in order to meet tourist satisfaction, to enhance tourist's experiences (more details are provided in Chap. 4), and to drive tourism towards sustainability [9].

1.3.2 Life Cycle Approach

Life cycle (LC) approach can be applied to any products or services to assess environmental impacts, life cycle cost, and social issues. The tools used to assess environmental impacts, life cycle cost, and social issues are life cycle assessment (LCA), life cycle costing (LCC), and social-life cycle assessment (S-LCA), respectively. Recently, there is another life cycle approach on sustainability issues called life cycle sustainability assessment (LCSA) [22].

Principles and framework of LCA were developed in 2006 and described in ISO 14040:2006 [21]. LCA consists of four main steps, which are goal and scope definition, life cycle inventory analysis (LCI), life cycle impact assessment (LCIA), and interpretation of the results. Among the four steps, the second step, LCI, is crucial since the quality of LCA results is reflected by the quality of inventory. Thus, LCA practitioners need good quality of data. However, currently, there is a lack of reliable databases to analyze tourism situation.

Each tourism component consumes inputs to generate outputs. Inputs are such as natural resources, cost, and labors. Outputs are, e.g., products (which is tourism, in this case a service product), emissions to the atmosphere, water, soil, noise pollution, and a range of social and cultural issues. Quantification of resource consumption and waste generation from supply-side is necessary in order to efficiently operate each tourism service. LCA is an applicable approach to quantify resource consumption and waste generation of each tourism service. More details on LCA application in tourism sector are presented in Chap. 2.

Basically, life cycle of each tourism component consists of four main phases, which are raw material acquisition, manufacturing, use, and end-of-life phases, as depicted in Fig. 1.3. For example, a hotel room requires inputs such as bed sheets, pillows, soap, shampoo, and other amenities, provides the service to customers, and generates solid wastes, wastewater, and emissions, which are needed to be disposed of. LCA approaches can be used to quantify areas where tourism activities are exceptionally burdensome to the environment, and improvements can be analyzed.

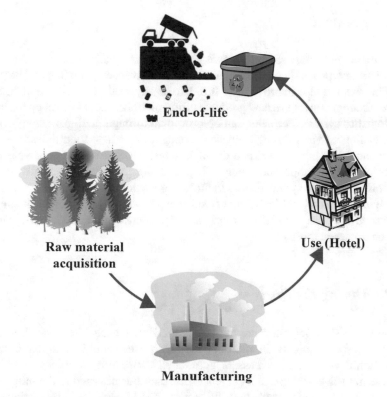

Fig. 1.3 Life cycle stages of a hotel

1.4   Tourism Enterprises During a Crisis to Avoid Unsustainable Tourism Activities

The coronavirus disease 2019 (COVID-19) pandemic, for instance, has caused economic crisis since all tourism services are halted to avoid the spreading of virus. Though ecosystems of many tourist destinations are revived during the pandemic, the COVID-19 may lead to several unsustainable tourism activities to prevent the spread of coronavirus. For example, more single-use plastics are used in food delivery services during the pandemic, resulting in a significant amount of solid wastes.

Future work from an integrated LCA-SCM framework perspective should focus on how to deal with such incident while maintaining sustainable development of the tourism sector. The work can be divided into three phases, which are pre-crisis, during the crisis, and post-crisis.

1.4.1 Pre-crisis

The tourism enterprises should prepare how to operate their businesses during the crisis. The enterprises should be prepared and had advanced planning, and be proactive. They should also improve their ICT skills, stay well-connected, and be informative. Examples are: (1) a new product selling platform should be set up to support their logistics system. The sales can be promoted through platform while avoiding the spreading of virus through physical contact; (2) a new food delivery service system should be more concerned on two issues, which are greenhouse gas emission from transportation and single-use packaging. An environmentally friendly food delivery service system should include, e.g., developing new packaging design and relying more on local food delivery service providers where food packaging can be returned and reused; and (3) a "tourism independence program" initiatives to rely more on domestic tourism.

1.4.2 During the Crisis

During the crisis when cities are put under lockdown, each enterprise should stay calm and respond quickly. In addition, the enterprises should try to maximize the use of their fixed assets. Fixed assets include, e.g., buildings, machines, and kitchen equipment. Each enterprise can: (1) reposition their business and re-defining market segments. For example, hotels with fully-equipped kitchen could reposition themselves as catering service providers. They can either sell foods or have their kitchen for rent. (2) communicate and collaborate with their partners, both in the same and different supply chains. For example, restaurants or bakeries with raw materials inventory could sell their raw materials rather than selling their finished goods to their network.

1.4.3 Post-crisis

After the crisis, the tourism sector should focus on, at least, three aspects, which are tourist's experience, financial issue, and organizational learning and knowledge management. For tourist's experience, sanitation and hygiene is the key for post-COVID tourism. Therefore, tourism-related service providers should pay attention on cleaning procedure. This would assure tourists that the place is clean and safe. For financial issue, government's financial aid should be allocated to local authorities to boost local economy, or set up fund for job creation. For organizational learning and knowledge management, enterprises should focus on: how to share lesson learned and best practices among the stakeholders of the entire supply chain of the tourism sector, and how educational institutes update and enrich their teaching contents with the case studies provided by the tourism and hospitality industry.

References

1. Forster, P., Ramaswamy, V., Artaxo, P., Berntsen, T., Betts, R., Fahey, D.W., Haywood, J., Lean, J., Lowe, D.C., Myhre, G., Nganga, J., Prinn, R., Raga, G., Schulz, M., Van Dorland, R.: Changes in atmospheric constituents and in radiative forcing. In: Solomon, S., Qin, D., Manning, M., Chen, Z., Marquis, M., Averyt, K.B., Tignor, M., Miller, H.L. (eds.) Climate change 2007: The Physical Science Basis. Contribution of Working Group I to the Fourth Assessment Report of the Intergovernmental Panel on Climate Change. Cambridge University Press, Cambridge (2007)
2. France-Presse, A.: Thailand Closes Dive Sites Over Coral Bleaching Crisis. (2016)
3. Alderman, L.: A World Without Ice? Iceland is Preparing. (2019)
4. Millan Lombrana, L., Warren, H., Rathi, A.: Measuring the Carbon-Dioxide Cost of Last Year's Worldwide Wildfires. (2020)
5. Scott, D., Hall, C.M., Gössling, S.: Tourism and Climate Change. Impacts, Adaptation and Mitigation. Routledge, London (2012)
6. Santarius, T., Walnum, H.J., Aall, C.: Rethinking Climate and Energy Policies: New Perspectives on the Rebound Phenomenon, pp. 209–225. Springer International Publishing, Cham (2016)
7. Aall, C., Hall, C.M., Groven, K.: Tourism: applying rebound theories and mechanisms to climate change mitigation and adaptation. In: Santarius, T., Walhum, H.J., Aall, C. (eds.) Rethinking Climate and Energy Policies. Springer, Cham (2016)
8. World Travel & Tourism Council: Travel & Tourism: Economic Impact 2019 World. (2019)
9. Zhang, X., Song, H., Huang, G.Q.: Tourism supply chain management: a new research agenda. Tour. Manag. 30(3), 345–358 (2009)
10. UN Environmental Programme: Responsible Industry: Tourism. [cited 2020 March 19]; Available from: https://www.unenvironment.org/explore-topics/resource-efficiency/what-we-do/responsible-industry/tourism
11. Gössling, S., Peeters, P.: Assessing tourism's global environmental impact 1900–2050. J. Sustain. Tour. 23(5), 639–659 (2015)
12. UNWTO: Sustainable Development. [cited 2020 March 19]; Available from: https://www.unwto.org/sustainable-development
13. Janusz, G.K., Bajdor, P.: Towards to sustainable tourism – framework, activities and dimensions. Procedia Econ. Finance. 6, 523–529 (2013)
14. UN World Tourism Organization: Tourism and the Sustainable Development Goals. UN World Tourism Organization, Madrid (2015)
15. Garnett, T.: Three perspectives on sustainable food security: efficiency, demand restraint, food system transformation. What role for life cycle assessment? J. Clean. Prod. 73, 10–18 (2014)
16. Adriana, B.: Environmental supply chain management in tourism: the case of large tour operators. J. Clean. Prod. 17(16), 1385–1392 (2009)
17. Arcese, G., Merli, R., Lucchetti, M.C.: Life cycle approach: a critical review in the tourism sector. In: The 3rd World Sustainability Forum (2013)
18. Budeanu, A., et al.: Sustainable tourism, progress, challenges and opportunities: an introduction. J. Clean. Prod. 111, 285–294 (2016)
19. Cerutti, A.K., et al.: Assessment methods for sustainable tourism declarations: the case of holiday farms. J. Clean. Prod. 111, 511–519 (2016)
20. De Camillis, C., Raggi, A., Petti, L.: Tourism LCA: state-of-the-art and perspectives. Int. J. Life Cycle Assess. 15(2), 148–155 (2010)
21. International Organization for Standardization: ISO 14040:2006 Environmental Management - Life Cycle Assessment - Principles and Framework. (2006)
22. UN Environmental Programme.: Towards a Life Cycle Sustainability Assessment: Making Informed Choices on Products. (2011)

23. Office of the National Economic and Social Development Board.: The Twelfth National Economic and Social Development Plan (2017–2021). In: Office of the Prime Minister. (ed.) Bangkok, Thailand
24. Pratomlek, O.: Critical Nature: Community-Based Tourism in Thailand: Impact and Recovery from the COVID-19. (2020)
25. Thai, P.B.S.: News เปิดปม: ตลาดควายนวัตวิถี. (2019)
26. Kozak, M., Martin, D.: Tourism life cycle and sustainability analysis: profit-focused strategies for mature destinations. Tour. Manag. **33**(1), 188–194 (2012)
27. TIES: The International Ecotourism Society. (2020) [cited 2020]; Available from: https://eco-tourism.org/
28. Fennell, D.A.: Ecotourism. Routledge, Abingdon (2020)
29. CREST: In: Hogenson, S. (ed.) The Case for Responsible Travel: Trends & Statistics 2017. Center for Responsible Travel, Washington, D.C. (2017)
30. Lee, W.H., Moscardo, G.: Understanding the impact of ecotourism resort experiences on tourists' environmental attitudes and behavioural intentions. J. Sustain. Tour. **13**(6), 546–565 (2005)
31. Buckley, R.: Evaluating the net effects of ecotourism on the environment: a framework, first assessment and future research. J. Sustain. Tour. **17**(6), 643–672 (2009)
32. Das, M., Chatterjee, B.: Ecotourism: a panacea or a predicament? Tour. Manag. Perspect. **14**, 3–16 (2015)
33. Nyaupane, G.P., Thapa, B.: Evaluation of ecotourism: a comparative assessment in the Annapurna conservation area project, Nepal. J. Ecotour. **3**(1), 20–45 (2004)
34. Horton, L.R.: Buying up nature: economic and social impacts of Costa Rica's ecotourism boom. Lat. Am. Perspect. **36**(3), 93–107 (2009)
35. Neto, F.: A new approach to sustainable tourism development: moving beyond environmental protection. In: Natural Resources Forum. Wiley Online Library (2003)
36. United Nations: In: Department for Economic and Social Information and Policy Analysis (ed.) Glossary of Environment Statistics, Studies in Methods. United Nations, New York (1997)

Chapter 2
The Role of Life Cycle Approaches in Sustainable Development of Tourism

2.1 An Introduction to Life Cycle Approaches

Life cycle approaches are a concept that considers a product or service over its entire life cycle. The life cycle of any products or services generally consists of four phases, which are raw material acquisition, use, manufacturing, and end-of-life [1]. Life cycle assessment or analysis (LCA) approaches have been widely used to quantify life-cycle environmental impacts and costs. There are three major benefits from evaluating a product or service from a life cycle perspective. The benefits are (1) to improve product or service performance, (2) to avoid unintended consequences, and (3) to aid in decision-making process.

1. *To improve product or service performance*
   Using life cycle approach can help identify a hotspot of a product or service. The hotspot can be the process with major cost, major resource consumption, or major environmental impact contribution. Once the hotspot process is identified, then the process can be improved by adopting technology or simply reducing unnecessary resource consumption to reduce costs or environmental impacts. For example, in operating a hotel, there are several sources of energy consumption such as air conditioners, lightings, televisions, and water heaters. From a life cycle perspective, sources of energy consumption are monitored to identify hotspots. The most energy and cost intensive process, which is generally a result of high quantity of resource consumed, in operating a hotel room is from using an air conditioner [2]. Therefore, thermal insulation materials or energy efficient air conditioners should be installed or a building with an effective air ventilation system should be designed to lessen the use of air conditioners.
2. *To avoid unintended consequences*
   Considering a product or service from a life cycle perspective can avoid any unintended consequences. Unintended consequences may occur as an unexpected outcome along with the main purpose. Such case can be seen from the shift from

© The Author(s) 2021
K. Soratana et al., *Supply Chain Management of Tourism Towards Sustainability*, SpringerBriefs in Environmental Science,
https://doi.org/10.1007/978-3-030-58225-8_2

fossil fuel to first-generation biofuels (from sugarcane and corn). The shift can mitigate global warming potential but promote eutrophication from fertilization during sugarcane and corn cultivation processes [3]. Similar case in the tourism sector is switching from single-use plastic to biomaterial containers. Switching from single-use plastic to biomaterial container, rather than multiple-use containers, e.g. glass or ceramic, results in lower solid wastes but higher greenhouse gas emissions (GHGs) due to methane (CH_4) generated from mismanagement of biomaterial wastes. Another example is overtourism, caused by tourism marketing or policies, e.g. top-down policy, without social involvement. The policy aims to boost economy of local community; however, a drastic change in number of visitors could cause several adverse impacts, e.g. a shortage of water resource, deterioration of natural areas, and interruption of local lifestyle. Therefore, a preliminary study should be conducted prior to any further development from all aspects. Life cycle approach is an appropriate tool for such cases.

3. *To aid in decision-making process*
   Making an effective decision requires data and information to support the process. One crucial step in life cycle approach is to construct a comprehensive list of inventories. Thus, life cycle approach is a quantitative method where options can be quantitatively compared. Decision-making can be achieved based on life cycle inventory. However, it should be noted that there are tradeoffs in any situation. Life cycle approach should be coupled with weighting techniques for a more specific condition of a particular tourism destination.

Based on the three main benefits of life cycle approach, the approach is applicable, thus should be adopted to manage issues in the tourism sector in all three pillars of sustainability. LCA has been applied to, e.g., package holidays, accommodations, impacts generated by one tourist, and short-haul tourism. Additional information on chronicle of LCA applications in tourism sector can be found in the study by [2].

2.2   Life Cycle Assessment and Its Implication

Life cycle assessment (LCA) is an appropriate tool to assist in balancing environmental pillar of the sustainability. As defined by the International Organization for Standardization (ISO) in the 14,040 series, LCA is a tool to quantify life-cycle resource consumption and environmental impact contribution from a product or service [1]. LCA can be conducted on products, processes, and even large systems. For LCA of products, inputs and outputs over products' life cycle are quantified and then converted into environmental impacts. The environmental impacts can be calculated based on emission conversion factors provided by, e.g., Intergovernmental Panel on Climate Change (IPCC). The life cycle generally includes raw material acquisition, manufacturing, use, and end-of-life phases of a product or service. LCA may refer to as a process from cradle to grave, cradle to gate, or cradle to cradle [4], as illustrated in Fig. 2.1.

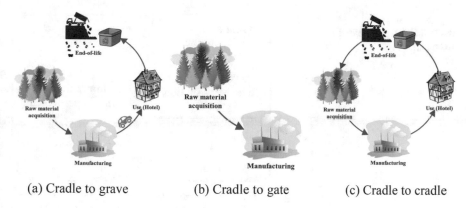

(a) Cradle to grave   (b) Cradle to gate   (c) Cradle to cradle

Fig. 2.1 Life cycle process of a product or service

A cradle-to-grave design includes all life-cycle phases; a cradle-to-gate design includes raw material acquisition and manufacturing of life-cycle phase; and a cradle-to-cradle design includes all life-cycle phases and the wastes generated will be re-circulated and used as a raw material in the manufacturing phase.

LCA according to ISO comprises four main steps, which are (1) goal and scope definition, (2) life cycle inventory analysis (LCI), (3) life cycle impact assessment (LCIA), and (4) interpretation of the results. Iterations between steps are common, for example, additional inventories might be required if more products and/or inputs were added to the system boundary defined in the goal and scope definition step, as depicted in Fig. 2.2. ISO's official reporting mechanisms for LCAs are called environmental product declarations (EPDs) and are defined by ISO 14025. Product category rules (PCRs) are documents that provide rules, requirements, and guidelines for developing LCAs and EPDs for specific products; they help ensure that LCAs are comparable and transparent.

2.2.1   Goal and Scope Definition

This step establishes and identifies objective(s), system boundary, and a functional unit of the LCA. The analysis can be a process LCA (only one product or service is evaluated) or a comparative LCA (more than one product or service are compared). System boundary defines processes, inputs (resources required/consumed), and outputs (emissions and wastes generated) included in the study. Examples of processes included in a tourism system are transportation to and from the destination, accommodation, restaurant, including other suppliers for businesses at the destination. A functional unit is used to quantitatively measure and fairly compare performance of products or services in relation to their inputs and outputs. Examples of the

Fig. 2.2 Life cycle
assessment steps (adapted
from [1])

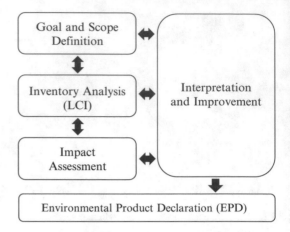

functional unit for tourism services are per one tourist per night, per one trip, or per one meal. This step of LCA will assist you in initiating a list of inventories to be collected. For a process with more than one product, a main product and co-product(s) should be identified, so inputs and impacts of the system can be allocated based on, e.g., mass or economic basis.

2.2.2   Life Cycle Inventory

LCI is a very critical step since the quality of LCA results depended mainly on the quality of LCI. LCI is generally a time-consuming step, particularly if the process you are investigating is new and/or unconventional, and there is no existing data/ database available for the process. Inventory data can be collected from various sources, either primary (e.g., experiment and questionnaire) or secondary (e.g., literature, government and corporate reports, patents, and life-cycle databases). Examples of life-cycle databases are *ecoinvent*, ETH-ESU, IdeMat, Franklin, and USLCI [5–9]. Inventories are collected according to the processes specified in the system boundary in order to achieve the defined goal in goal and scope definition step. Numerous life-cycle databases are provided by, e.g., academic institutes, government sectors, and consulting companies.

Ecoinvent The database has been developed by the Swiss Centre for Life Cycle Inventories. This database contains 4000 datasets from different LCI databases. The database includes products and services from the energy, transport, building materials, chemicals, pulp and paper, waste treatment, and agriculture mainly investigated under Swiss and Western European conditions.

USLC This database has been developed based on the US Input–Output (IO) database consisting of data from approximately 500 × 500 industrial sectors since 1998.

This matrix was then connected to a large environmental matrix from several environmental data sources, such as Toxic Releases Inventory 98 (TRI), Air Quality Planning and Standard (AIRS) data of the US EPA, and Energy Information Administration (EIA) data of the US Department of Energy. Later, the impact from small- and medium-sized enterprises (SMEs) and non-point sources such as transport have been added to the environmental-related matrix.

Tourism services, such as accommodations, restaurants, or transportation, enterprises could construct an inventory table by themselves. Inventories such as water and energy (gasoline, diesel, or electricity) could be monitored and collected from water meter and electric meter of a hotel room, or odometer and fuel efficiency of a vehicle. Other information and data required to calculate for total resource consumption over its life cycle are included. Some LCA studies end the analysis process in this step and obtain a footprint of water, energy, or carbon. Over the life cycle of a tourism service considered, major resource consumption process is revealed, thus enterprises could act on resource consumption reduction.

2.2.3   Life Cycle Impact Assessment

LCIA step is where the LCI data is converted and presented in terms of understandable and quantifiable environmental impacts, e.g. global warming, eutrophication, acidification, ozone depletion, and smog formation potentials. Three steps in conducting the LCIA include impact category definition, classification, and characterization. Examples of LCIA tools are, e.g., CML 2001, Eco-Indicator 99, EDIP 2003, IMPACT 2002+, and TRACI (Tool for Reduction and Assessment of Chemical and other environmental Impacts).

TRACI is one of the most common LCIA tools used in the USA developed by the US EPA, particularly for the USA. It provides ten environmental impact categories, which are ozone depletion, global warming, smog, acidification, eutrophication, carcinogenic, non-carcinogenic, respiratory effects, ecotoxicity, and fossil fuel depletion potentials. Each TRACI impact is calculated on a midpoint basis to avoid an estimation or a forecasting in the LCIA while still can reflect the stressors and the effects of the contaminants [10, 11]. Midpoint characterizations are presented in kg equivalent (eq) of a reference substance; for instance, chlorofluorocarbon-11 (CFC-11) is a reference substance for ozone depletion potential, nitrogen for eutrophication potential, and CO_2 for global warming potential. It should be noted that different LCIA methods use different reference substance, thus resulting in different units and values. Midpoint impact categories can be allocated into one or more damage categories or endpoint. An LCIA tool IMPACT 2002+ allocates midpoint impact categories into four endpoint damage categories (human health, ecosystem quality, climate change, and resources), which allows different impact categories to be directly compared [12]. Another LCIA tool that uses damage-oriented approach is the Building for Environmental and Economic Sustainability (BEES). The BEES tool combines different impact category performance by multi-attribute decision analysis and defines the relative contribution of each impact category to the environment by weighting [13]. The LCIA results can be normalized, grouped, weighed, and analyzed to improve quality of the results.

Global warming potential can be manually calculated using GHGs constituents conversion factor provided by IPCC. For example, based on an assumption that a hotel room consumes 14 kWh of electricity and each room is occupied for maximum of 22 h, CO_2 emission from the production of electricity is 0.448 kg CO_2 eq per kWh. Therefore, global warming potential from the electricity consumed during a stay in a hotel room is 14 kWh × 22 h × 0.448 CO_2 eq/kWh equals 138 kg CO_2 eq. The conversion factors of each GHG are listed in Table 2.1.

LCA software, such as SimaPro or OpenLCA, is available to aid in LCI and LCIA steps. SimaPro contains databases of inventories for materials and processes in several industries and LCIA tools.

2.2.4 Interpretation and Improvement Analysis

Interpretation and Improvement Analysis is where the LCI and LCIA results are correlated, interpreted, and improved to present meaningful information and to enable decision-making consistent with the defined goal and scope. Interpretation should deliver results and explain limitation to inform industries and decision maker [15–17]. In this step, emission hotspots and the impacts of a product that can be easily mitigated should be identified to provide a reduction of the life cycle impacts of the product.

The communication elements of LCA should be focused. For instance, the unit of global warming potential, i.e. kg CO_2 eq is not commonly known by general audiences. Therefore, when reporting global warming potential impact, it should be compared to more familiar resources, e.g. the number of tree needed to be planted or the distance of automobile traveling. For example, rather than reporting a global warming potential from occupying a hotel room for a day as 138 kg CO_2 eq, global warming potential should be reported as the number of trees required to sequester

Table 2.1 Summary of 20- and 100-year global warming potential value of GHG constituents [14]

GHGs	Chemical formula	Global warming potential for designated time horizon (kg CO_2 eq)	
		20 years	100 years
Carbon dioxide	CO_2	1	1
Methane	CH_4	72	25
Nitrous oxide	N_2O	289	298
Hydrofluorocarbons	HFCs	437–12,000	124–14,800
Perfluorocarbons	PFCs	5210–8630	7390–12,200
Chlorofluorocarbons	CFCs	6540–11,000	6130–10,900

the amount of CO_2. Assuming a mature tree can consume approximately 22 kg CO_2 eq per year or 0.05965 kg CO_2 per day, therefore, it would require to plant up to 2313 trees in order to mitigate 138 kg CO_2 eq. Another interpretation can be calculated based on inventory result. The energy consumption of 308 kW per room per day (14 kWh/room × 22 h/day) equals 77 charges of cell phone (which consumes approximately 2–6 Watts when charging). Infographic is also useful to deliver the message.

2.3   Tourism Activities Over Their Life Cycle

This section aims to provide examples and details on how to conduct LCA and/or LCC. Explanations on how to draw a system boundary, define a functional unit, and construct inventories of each tourism activity in tourism supply chain are provided. Existing studies on LCA in tourism sector for tourism activities such as accommodations, restaurants, transportation, tour operators, and waste disposal facilities are reviewed. Generally, functional units used in the tourism sector are, e.g., environmental loads, resource consumed or cost per tourist per night, or per tourist per trip. The indicators used to present environmental loads from tourism are, e.g., petroleum fuels, electricity and water consumption, solid waste and wastewater discharges, and CO_2 emission over the life cycle of a tourism service or process [18].

In dealing with a tourism service or process with multiple products (e.g., main products, by-products, and co-products), there are two approaches. The two approaches are (1) attributional LCA and (2) consequential LCA [19]. *Attributional LCA* is suitable for an isolated system where all impacts are attributed to each of the system products by allocation or displacement. The method presents impacts from the entire life cycle of the product/service and its sub-systems. On the other hand, *consequential LCA* focuses on the effects of changes made within the life cycle [20]. The method can also be called as a change-oriented or prospective LCA. It considers a much broader system boundary and normally deals with co-products by system expansion. Consequential LCA is also applicable to prospectively indicated impacts from policy [21].

In the case where the production system results in several products, one product is usually defined as a primary product, while the other remaining products are defined as co-products. The original system boundary is then expanded to include the production of a displaced co-product. The avoided environmental impacts from the production of the co-products are credited to the primary product. Allocation of primary product and co-products can be based on economic, mass, or energy values, depending on the products investigated. For example, assuming a system boundary of serving a cup of coffee (main product) at a coffee shop, where its ground coffee (co-product) is used as a fertilizer. Using system expansion, environmental impact mitigation from using ground coffee as a fertilizer should be deducted from the total environmental load.

At the time of writing, there were no product category rules (PCRs) for sectors related to tourism (e.g., hotels, restaurants, general tourism). There are, however, PCRs for some forms of transportation (freight transport of goods) and fuels, but not airplanes. PCRs are kept up to date online and can be found at www.environdec.com/PCR.

Constructing an LCA of tourism activities is no easy task. Existing LCA tools and databases typically have data for unit processes, e.g. a plate, a towel, a boiler used in a hotel, a bus used to transport people. There are no datasets that comprise entire hotel operations or restaurant operations, for example. The data required to construct a LCA for an entire tourism industry, including all hotel operations, tours, and travel would be extensive.

2.3.1   Accommodations

Accommodation is one of the three principle elements in tourism, in addition to transportation and recreation activities [18]. Entrepreneurs should start conducting a LCA by specifying whether their businesses are considered as a hotel, bed, and breakfast, a campground, or a homestay. Hence, kind of destination, type, operations, and processes of the accommodation can be defined accordingly. Goal and scope should be defined and also reflected in a system boundary. Processes include in the system boundary are services provided at accommodations, e.g. guest rooms, laundry, meals, and swimming pools, based on the goal of the analysis. Resource consumption of each process is accounted from arrival of tourists at the hotel to departure of tourists [22]. Environmental impacts and cost of the construction of the hotel building may be cut off from the system boundary [23]. Functional unit can be environmental loads or cost *per one guest night*. Additional services such as breakfast and laundry may be included. Though the analysis is on accommodation, inventories may vary, depending on the goal and scope. Inventories consist of information and data related to energy, water, solid wastes, and other resources. For example, if water resource is the major spatial concern where the hotel is located, you should conduct a water footprint analysis on your hotel operation to investigate for a water hotspot. Based on literature review, the major environmental impacts occurring from accommodations are water depletion and acidification due to services of guest rooms and laundry [22]. Complex life cycle inventories of the quantity of water consumed by accommodation can simply begin with the information provided in water bill. Life cycle inventories should include activities related to water production and consumption, and quantity of water used by each activity, as listed in Table 2.2.

Table 2.2 Example of life cycle inventory for a water footprint of one guest for one night stay in a hotel room [by Kullapa Soratana]

converted to m^3 per functional unit

e.g. from primary or secondary sources

Process/ Activity	Quantity (unit)	Source
Hotel room:		
No. of shower	2/day	
Shower	0.065 m^3/shower	
No. of bath	1/day	
Bath	0.136 m^3/bath	
No. of toilet use	5/day	
Toilet	0.006 m^3/flush	
Size of a bathroom	2.0 m × 1.7 m	
Swimming pool:		
Pool size	145 m^3	
Evaporation rate	4.65 mm/day	
Laundry:		
No. of washing machine	5	
Washing machine (front loading)	0.023 m^3/kg of dry clothing	
Washing load	8 kg/load	

2.3.2   Transportation

Transportation is another principle element in tourism [18]. Generally, air transport contributes the highest emissions, such as CO_2, CO, hydrocarbon, and NO_x, since it consumes the largest quantity of fuel, compared to other modes of transport [18]. Other types of vehicles used in tourism are, e.g., motorcycles, rental cars, tour buses, and small shuttle buses. Each vehicle has its purpose either for domestic or international transport. Each trip may require various modes of transport for visitors to travel from the origins to their destinations. Therefore, it is essential to identify the origin and the destination and conventional modes of transport of a trip. For example, it is common for a hotel to use a coach as a shuttle to transport visitors to and from an airport. Besides, a trip from a mainland to an island would require a visitor to take an airplane, a coach, a ship, and a rental car to arrive at the destination, as illustrated in Fig. 2.3. Therefore, several modes of transport can be included in a system boundary.

   In conducting a LCA on transport services in the tourism sector, factors to be considered are, e.g., types of vehicles, distances, fuel efficiency, fuel types, and passenger occupancy rate. Inventories on the factors can be primarily collected from the enterprise and from secondary data sources, e.g. government reports and literature. System boundary is defined based on the goal and scope of the LCA. The study can be *a process LCA* where one mode of transport is investigated for a hotspot. The

Fig. 2.3 System boundary of transportation from origin to island destination

hotspot is the major impact contribution, the most cost intensive process or a room for process improvement. After hotspot identification, a reduction on energy and other resource consumptions and cost can be made. Another can be *a comparative LCA* where several modes of transport are compared for the most appropriate mode.

It should be noted that, generally, there are tradeoffs among transportation options. For example, one of the environmental performance improvements of the vehicle is to switch from using fossil fuel to biofuel or electricity. Both biofuel and electric vehicles are preferred since they emit less emission during the use phase comparing to a conventional fossil fuel vehicle. However, biofuel vehicles and electric vehicles (EVs) have their drawbacks. The production phase of biofuels involved with an agricultural process which consumes fertilizers and other sources of energy. The agricultural process causes eutrophication impact potential from fertilizer application. For EVs, the end-of-life phase of the vehicles should be concerned. A method to manage obsolete EV batteries, in addition to landfilling, should be explored. Weighting methods can be coupled with LCA in selecting the most suitable option under a certain circumstance.

There are several functional units used in the transportation sector, depending on the goal and scope of a LCA. One of the common functional units used in transportation is *per ton-km*. The functional unit equals to the transportation of one metric ton (1000 kg) of freight or passenger for 1 km. Other functional units used in the transportation sector are *energy use per tourist per trip*, *the environmental loads of all tourists visited one hotel in its seasonal operation*, [24]. Then, LCIA results can be reported as a quantity of energy consumed or CO_2 eq per guest per trip. From a supply chain and LCA perspectives, environmental performance of tourism supply chain can be improved. Stakeholders, e.g. hoteliers or tour operators may suggest or provide the most sustainable transportation modes or routes in traveling to and from the destination.

2.3.3 Restaurants

The restaurant industry is considered the least world sustainable economic sectors based on the green GDP [25]. Restaurants contribute significant amount of greenhouse gas emissions due to their energy intensive process [25]. Waste generation is another issue of restaurant services. General practices for restaurants to minimize solid waste are to avoid using single-serving condiment and sauce sachets, avoid using single-use plastic or Styrofoam containers and straws or to provide them only upon request. Food waste from restaurants can be minimized by preparing appropriate food portion, while wastewater can be managed by removing grease and oil before drainage [26]. Often, all these practices cannot be simultaneously implemented. Therefore, resource, environmental impacts, and costs of these practices should be quantified to aid in practices prioritization and tradeoffs identification among the practices.

A LCA of a restaurant can be a farm-to-fork analysis or a farm-to-farm gate analysis. For a farm-to-fork analysis, system boundary consists of four processes, which are food procurement, food storage, food preparation and cooking, and food service and operational support [27]. Functional unit of a restaurant can be *per customer*, *per service*, or *per meal*. LCI can be developed according to the activities and resource consumed in each process. Inventories can be collected from meters, water and electricity bills, or purchasing order. For LCIA, environmental impacts of resource consumed in the processes can be calculated based on inventories in databases, e.g. ecoinvent, USLCI, and peer-reviewed publication. For example, greenhouse gas emissions of on-farm produce and livestock are available in the study by Gossling et al. [28]. There is a range of paid LCIA tools available, e.g. CML 2001, Eco-Indicator 99, EDIP 2003, IMPACT 2002+, and TRACI.

2.3.4   Tour Operators

Tour operators play a vital role in promoting sustainability. Due to tour operators' multi-faceted functions to stay in business, they have potential to enhance sustainability throughout tourism supply chain. In addition, many major mass tourism countries, e.g. Spain, Greece, and Italy, depend on tour operators to persuade tourists to use the countries tourism facilities [29]. Thus, tour operators can act as a screening platform to select sustainable tourism services to be used in a holiday package. Among various options of services in tourism sector, tour operators could adopt LCA to distinguish between sustainable service and "greenwashing" service. Therefore, tour operators could adopt LCA [23] to:

1. Evaluate environmental performance.
2. Identify hotspots for internal purposes or environmental standard compliance.
3. Compare different environmental innovation/technologies.
4. Support environmentally friendly services and green market.

Life cycle thinking was first applied to the tourism sector on the tour operator in UK since 1998 [22]. The study examined environmental impacts of major destinations and the management of tourist facilities of the all-inclusive package holiday. The goals were to identify environmental hotspots and define improvement actions for each stakeholder in the supply chain [23]. Currently, there is a sustainability management system called Travelife providing, e.g., training, certification, and practical solutions as a guidance to tour operators in moving towards sustainability [30]. This renowned system could also be improved by incorporating LCA along with their other services.

Sustainable tourism services can also be evaluated by using criteria. One of the well-known criteria is developed by the Global Sustainable Tourism Council (GSTC) for tour operators, hotels, and destinations. The establishment of the criteria focuses on four aspects, which are (1) effective sustainability planning, (2) social and economic benefits maximization for the local community, (3) cultural heritage enhancement, and (4) negative environmental impact reduction [31]. However, the criteria cannot be used to quantify resource consumption, environmental impacts, and costs but to guide the sector in developing sustainability by following the standards. The well-established GSTC criteria can be integrated with life cycle thinking to drive sustainability in the tourism industry, which becomes quantifiable. Hence, the integration avoids greenwashing businesses. Criteria based on an integration of the GSTC criteria for destinations and life cycle thinking are illustrated in Table 2.3. The indicators can be modified according to the specification of each destination and country.

Table 2.3 Criteria from the integration between GSTC criteria for destinations and life cycle thinking

Activities	Stages	Criteria	Score	Indicators
Destination	1.1 Raw material	1.1.1 Appropriateness of attraction to local environmental conditions	0	Inappropriate - no accessibility, no safety, not good for local environmental conditions
			1	Accessible
			2	Accessible and safe for tourists
			3	Accessible, safe for tourists and good for local environmental conditions
		1.1.2 Quality of local's environment	0	Deteriorate
			1	As good as normal condition
			2	Has been improved to normal condition
			3	Better than normal condition
	1.2 Operation	1.2.1 Budget for environmental management	0	None
			1	Inadequate amount of budget
			2	Sufficient amount of budget, but has not been implemented as planned
			3	Sufficient amount of budget, and has been implemented as planned
		1.2.2 Units for environmental protection, including cleaning	0	None
			1	No direct responsible units
			2	Insufficient direct responsible units
			3	Sufficient direct responsible units
		1.2.3 Systems to manage, control and monitor resource consumption	0	None
			1	There is management system
			2	There are management and controling systems
			3	There are management, controling and monitoring systems
		1.2.4 Natural disaster prevention	0	None
			1	There are prevention, mitigation and preparedness plans
			2	There are prevention, mitigation, preparedness and response and relief plans
			3	There are prevention, mitigation, preparedness, response and relief and rehabilitation and reconstruction plans
		1.2.5 Communication on environmental awareness for tourists	0	None
			1	Posters to promote environmental awareness
			2	Projects to promote environmental awareness
			3	Projects with tourists' involvement to promote environmental awareness
		1.2.6 Policy, laws and regulations related to attraction's environment	0	None
			1	There are policies, laws and regulations, but not clear implementation plans
			2	There are policies, laws and regulations with clear implementation plans
			3	There are successful projects from the policies, laws and regulations
	1.3 Use	1.3.1 Promoting on energy consumption reduction to stakeholders	0	None
			1	Posters to promote awareness on energy consumption reduction
			2	Projects to promote awareness on energy consumption reduction
			3	Projects with tourists' involvement to promote the awareness
		1.3.2 Promoting on water consumption reduction to stakeholders	0	None
			1	Posters to promote awareness on water consumption reduction
			2	Projects to promote awareness on water consumption reduction
			3	Projects with tourists' involvement to promote awareness on water consumption reduction
		1.3.3 Promoting on other resources consumption reduction to stakeholders	0	None
			1	Posters to promote awareness on other resources consumption reduction
			2	Projects to promote awareness on other resources consumption reduction
			3	Projects with tourists' involvement to promote the awareness
	1.4 End-of-life	1.4.1 Wastewater treatment system before discharging	0	None
			1	There is wastewater treatment system, but does not meet the standard
			2	There is a standard wastewater treatment system
			3	There is a wastewater treatment system which is better than the standard
		1.4.2 Waste disposal system, including waste collection system	0	None
			1	There are waste collection points, but no decent waste collection system
			2	There are waste collection points and system, but no decent waste disposal system
			3	There are decent waste collection points, collection system and waste disposal system
		Total	**0**	

2.3.5   Waste Disposal Facilities

Another process that we should pay attention to is waste disposal in the tourism industry. According to waste management hierarchy, the most preferable method is prevention. However, sometimes, it is unfeasible to operate tourism activities without using any resource. The next most preferable methods are 3Rs, consisting of reduce, reuse, and recycle, respectively [32]. 3Rs methods can be accomplished through an effective waste separation system. In addition, an existence of waste recycling and composting facilities is also important. Various types of wastes from processes in tourism are, e.g., solid wastes either hazardous or non-hazardous, food wastes, wastewater, and other emissions. Each type of wastes can be dealt with differently. For example, solid wastes that are not reusable or recyclable should be disposed of by landfilling, combustion, or incineration. Though, evidently, landfill is the worst waste management method in most of the European cities [33]. Therefore, organizations are aiming for zero landfill. Many national parks or recreation islands are aiming for zero waste by not allowing any waste left in the parks. All wastes must be disposed of outside the parks [34]. However, this management only focuses on the recreation site; waste management system should be considered over the life cycle of waste.

Before defining goal and scope, sources and types of wastes must be identified. The selected or all possible waste disposal methods are considered. System boundary may include from waste collection, sorting, disposal (e.g., recycling, landfilling, and combustion), and co-products (e.g., methane gas, electricity, and biofertilizer). Quantity and other related information on wastes must be collected in order to find the most appropriate waste disposal method. If resource consumption such as energy and water are the major concerns, then the investigation can be finished at LCI step of LCA. LCI is crucial and time consuming. The accuracy and reliability of the LCIA results depend largely on LCI. Moreover, a comparative LCA can be conducted to quantify cost, environmental impacts, and tradeoffs. The results can be used to support decision-making process; mismanagement of waste can be prevented.

2.4   Good Practices in Using LCA to Promote Sustainable Development Tourism

Before any further discussion, it should be mentioned that life cycle approach is not a solution to improve environmental performance or to reduce cost itself, but a tool to identify hotspot and tradeoffs. Therefore, good practices in this case are the use of LCA in different tourism sectors to drive tourism towards sustainability; in another word, to keep the three pillars of sustainability in balance. Life cycle approach can be applied throughout the life cycle of products and services in the tourism sectors, as illustrated in Fig. 2.4.

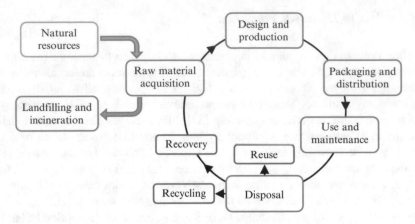

Fig. 2.4 Life cycle approach for tourism product. (Adapted from)

For instance, life cycle approach can be applied to a management scheme in the restaurant sector. The management follows three aspects of the green restaurant standards which are foods, environment and equipment, and management and social responsibility [25].

1. Foods—procurement, menu planning and cooking, and take-out packaging.

Raw Material Acquisition   Green restaurants generally opt for organic food and/or local food for low carbon menu. *Organic food* comes from organic farming, which uses organic fertilizer and natural pest control, while maintaining biodiversity of the ecosystem. *Local food* is food prepared from produce in vicinity. From life cycle perspective, both options do not necessarily always benefit the environment or cost. For example, organic produce tends to benefit the environment with high production cost and adverse environmental impact from tailpipe emissions, if the produce is not locally grown. For local produce, the environmental benefit may come from a short-distance shipping, however, the produce should be indigenous plants to avoid high maintenance cost and environmental impacts from, e.g., fertilizers and energy consumption for greenhouse environmental control system.

Packaging and Disposal   For take-out packaging, there is a wide range of materials from plastics, paper, and eco packaging. The disposal phase of packaging should be taken into consideration. Some plastics food containers are recyclable, while eco packaging is compostable. However, this is on the condition that a recycling facility and a composting system exist. Therefore, stakeholders in the restaurant industry should adopt and conduct a process and a comparative LCA to quantify environmental impacts over the life cycle of organic and local foods, and different types of packaging.

2. Environment and equipment—environment of kitchen and dining area, and cleaning and post-treatment.

Design and Production Kitchen and dining of a sustainable restaurant should be equipped with energy efficient and water efficient appliances. Natural light and air ventilation should be employed to reduce electricity consumption. Detergents, either for dishes or kitchen equipment, used in the restaurant should be eco-friendly. Therefore, gray water can be collected for watering plants. Besides, other wastes such as fat, oil, and grease, and hazardous waste (e.g., lamps and batteries) must be handled appropriately, while food waste must be composted. Apparently, kitchen and dining areas consume energy and chemicals (detergents); several options of appliances and chemicals are available. Thus, LCA is an appropriate tool to aid in decision-making process before purchasing.

3. Management and social responsibility—management policy, customer education, and corporate social responsibility (CSR)

Design and Use Management level has significant contribution to the success of sustainable restaurants by including activities to promote sustainability in the management policy. The activities are, e.g., green supplier selection, staff training program, and water and energy monitoring system. In addition, green building design concept and green building materials should be employed in restaurant construction. The green building concept considers environmental protection, resource conservation, and greenhouse gas reduction [32]. Using green building materials enhances the potential to meet the green building certification since the materials save energy and water resources. Examples of green building standards are LEED (Leadership in energy and environmental design) and BREEAM (Building research establishment environmental assessment method).

Customer education and CSR can be implemented by launching a program to reward customers who bring their own containers or putting up a poster to promote green behavior. Restaurants can reach out to the neighborhood communities for social involvement through green activities. Moreover, restaurants should reach out to their suppliers for collaboration on, e.g., providing green foods, using green transportation, and recycling wastes. Then, life cycle approach can be implemented to evaluate environmental impact and cost of products and/or services from suppliers.

In this aspect of the green restaurant standard, comparative LCA can be conducted on the status quo and after implementing the green programs (e.g., staff training, customer education, or marketing promotion). The results indicate an effectiveness of environmental and cost of the program. Thus, the results could support management level in decision-making process whether to continue, modify, or abort the programs.

References

1. International Organization for Standardization: ISO 14040:2006 Environmental Management - Life Cycle Assessment - Principles and Framework. (2006)
2. Michailidou, A.V., et al.: Life cycle thinking used for assessing the environmental impacts of tourism activity for a Greek tourism destination. J. Clean. Prod. **111**, 499–510 (2016)

3. Soratana, K., Khanna, V., Landis, A.E.: Re-envisioning the renewable fuel standard to mini-mize unintended consequences: a comparison of microalgal diesel with other biodiesels. Appl. Energy. **112**(Supplement C), 194–204 (2013)
4. McDonough, W., Braungart, M.: Cradle To Cradle: Remaking The Way We Make Things. North Point Press, New York (2002)
5. Spriensma, R.: SimaPro Database Manual, The BUWAL 250 Library. PRé Consultants, Amersfoort (2004)
6. Frischknecht, R., et al.: Ecoinvent Report No. 2, Code of Practice Data v2.0. Swiss Federal Institute of Technology Zürich (ETHZ), Dübendorf (2007)
7. Frischknecht, R., Jungbluth, N.: SimaPro Database Manual, The ETH-ESU 96 Libraries Version 2.1. ESU-Services, Schaffhausen (2004)
8. Norris, G.A.: SimaPro Database Manual, The Franklin US LCI Library Version 2.0. Sylvatica, North Charleston (2003)
9. Delft University of Technology: IdeMat Online. (2001)
10. Bare, J.C., et al.: TRACI: the tool for the reduction and assessment of chemical and other environmental impacts. J. Ind. Ecol. **6**(3–4), 49–78 (2003)
11. Bare, J., Gloria, T., Norris, G.: Development of the method and U.S. normalization database for life cycle impact assessment and sustainability metrics. Environ. Sci. Technol. **40**(16), 5108–5115 (2006)
12. Jolliet, O., et al.: IMPACT 2002+: a new life cycle impact assessment methodology. Int. J. Life Cycle Assess. **8**(6), 324–330 (2003)
13. Lippiatt, B.C.: Selecting cost-effective green building products: BEES approach. J. Constr. Eng. Manag. **125**(6), 448–455 (1999)
14. Forster, P., Ramaswamy, V., Artaxo, P., Berntsen, T., Betts, R., Fahey, D.W., Haywood, J., Lean, J., Lowe, D.C., Myhre, G., Nganga, J., Prinn, R., Raga, G., Schulz, M., Van Dorland, R.: Changes in atmospheric constituents and in Radiative forcing. In: Solomon, S., Qin, D., Manning, M., Chen, Z., Marquis, M., Averyt, K.B., Tignor, M., Miller, H.L. (eds.) Climate Change 2007: The Physical Science Basis. Contribution of Working Group I to the Fourth Assessment Report of the Intergovernmental Panel on Climate Change. Cambridge University Press, Cambridge (2007)
15. International Organization for Standardization: ISO 14040:2006, Environmental Management - Life Cycle Assessment - Principles and Framework, vol. 20. International Organization for Standardization, Geneva (2006)
16. Udo de Haes, H.: How to approach land use in LCIA or, how to avoid the Cinderella effect? Int. J. Life Cycle Assess. **11**(4), 219–221 (2006)
17. Udo de Haes, H., van Rooijen, M.: In: Edition, F. (ed.) Life Cycle Approaches: The Road from Analysis to Practice. UNEP/SETAC Life Cycle Initiative, Nairobi/Pensacola (2005)
18. Kuo, N.-W., Chen, P.-H.: Quantifying energy use, carbon dioxide emission, and other environmental loads from island tourism based on a life cycle assessment approach. J. Clean. Prod. **17**(15), 1324–1330 (2009)
19. Thomassen, M., et al.: Attributional and consequential LCA of milk production. Int. J. Life Cycle Assess. **13**(4), 339–349 (2008)
20. Ekvall, T., Weidema, B.P.: System boundaries and input data in consequential life cycle inventory analysis. Int. J. Life Cycle Assess. **9**(3), 161–171 (2004)
21. Kaufman, A.S., et al.: Applying life-cycle assessment to low carbon fuel standards: how allocation choices influence carbon intensity for renewable transportation fuels. Energy Policy. **38**(9), 5229–5241 (2010)
22. Arcese, G., Merli, R., Lucchetti, M.C.: Life cycle approach: a critical review in the tourism sector. In: The 3rd World Sustainability Forum (2013)
23. De Camillis, C., Raggi, A., Petti, L.: Tourism LCA: state-of-the-art and perspectives. Int. J. Life Cycle Assess. **15**(2), 148–155 (2010)
24. Michailidou, A.V., Vlachokostas, C., Moussiopoulos, N.: A methodology to assess the overall environmental pressure attributed to tourism areas: a combined approach for typical all-sized hotels in Chalkidiki, Greece. Ecol. Indic. **50**, 108–119 (2015)

25. Wang, Y.-F., et al.: Developing green management standards for restaurants: an application of green supply chain management. Int. J. Hosp. Manag. **34**, 263–273 (2013)
26. Environmental Research Institute of Chulalongkorn University and Bumi Kita Foundation: Sustainable Tourism Management in Thailand: A Good Practices Guide for SMEs. Environmental Research Institute of Chulalongkorn University and Bumi Kita Foundation, Bangkok (2007)
27. Baldwin, C., Wilberforce, N., Kapur, A.: Restaurant and food service life cycle assessment and development of a sustainability standard. Int. J. Life Cycle Assess. **16**(1), 40–49 (2011)
28. Gössling, S., et al.: Food management in tourism: reducing tourism's carbon 'foodprint'. Tour. Manag. **32**(3), 534–543 (2011)
29. Carey, S., Gountas, Y., Gilbert, D.: Tour operators and destination sustainability. Tour. Manag. **18**(7), 425–431 (1997)
30. Travelife: Tour Operators and Travel Agents: Training. Available from: https://www.travelife. info/index_new.php?menu=home&lang=en
31. Global Sustainable Tourism Council: GSTC Tour Operator Criteria. (2016)
32. Environmental Protection Agency: Sustainable Materials Management: Materials Management and the 3Rs Initiative. (2007); Available from: https://archive.epa.gov/oswer/international/ web/html/ndpm-3rs-initiative-and-materials-management.html
33. Cherubini, F., Bargigli, S., Ulgiati, S.: Life cycle assessment (LCA) of waste management strategies: landfilling, sorting plant and incineration. Energy. **34**(12), 2116–2123 (2009)
34. Przydatek, G.: Waste management in selected national parks: a review. J. Ecol. Eng. **20**(4), 14–22 (2019)

Chapter 3
A Knowledge Supply Chain for the ICT-Enhanced Tourism Education

3.1 ICT Usage in the Tourism Sector

Tourism has taken advantage of Information and Communication Technologies (ICTs). ICT has been using as a tool to bridge tourism suppliers, intermediaries as well as end-consumers, while simultaneously reengineering the industry towards increased customer satisfaction [1]. ICTs have been applied to the tourism industry in Europe and America for more than 40 years; however, the e-revolution in the Asian tourism sector has been piecemeal and ad hoc. In the Greater Mekong Sub-region (GMS), the tourism industry in these countries is facing serious challenges, such as an acute shortage of desired infrastructure and skilled human resources [2]. The result is that tourism Revenue Leakage (TRL) occurs, causing tourism revenue earned within a country to be "leaked" to foreign companies and tour operators [3]. This situation underlines the need to spread e-tourism education and raise awareness among GMS countries, where tourism education is provided mainly through higher education (HE) institutes, particularly in serving high technology and knowledge intensive industries like e-tourism.

This chapter focuses on how to solve the issues associated with e-tourism education through a knowledge engineering approach to identify and supply the knowledge for a curriculum. The structure of Chap. 3 begins with a background introduction on knowledge engineering (KE), knowledge management (KM), and knowledge supply chain (KSC), which is an integration of supply chain management (SCM) and KM. Then the methods on how to close the gap between knowledge supply and knowledge demand in e-tourism curriculum design are presented as follows: data collection and structuring, benchmarking, understanding the "as-is" situation, simulating the desired "to-be" situation. Finally, solutions to the two major problems in previous e-tourism curriculum research, namely, the long lead-time of the curriculum, the lack of knowledge provision and sharing, are identified.

© The Author(s) 2021
K. Soratana et al., *Supply Chain Management of Tourism Towards Sustainability*, SpringerBriefs in Environmental Science,
https://doi.org/10.1007/978-3-030-58225-8_3

3.2   KSC: The First Time to Integrate KM and SCM in the Context of Tourism and Its Education

3.2.1   KE, SCM, and Associated Tools

Knowledge Engineering (KE) is a conceptual approach to modeling knowledge in the form of computational constructs and software implementation [4]. While supply chain concepts and principles were developed in the manufacturing sector, they have been applied to the education sector [5]. Analogous to the production of goods in the manufacturing industry, products of higher education are intellectual assets, and higher education institutes are attempting to satisfy their stakeholders through a variety of methods. Although there is controversy about whether knowledge can be shifted and moved through a supply chain akin to physical products, the Knowledge Engineering (KE) perspective taken in this article considers that knowledge can be modeled to solve problems. Specifically, the research applies the Supply Chain Operations Reference (SCOR) model to capture the problems and processes of curriculum design.

3.2.2   KM and Its Tools

Knowledge Management (KM) is believed to be used by higher education institutes to gain a more comprehensive, integrative, and reflexive understanding of the impact of information on their organizations [6]. Researchers such as [7] elaborate the opportunities and practices for colleges and universities to apply knowledge management to support every part of their objectives. In terms of curriculum, they showed that KM can bring benefits to curriculum design and development in several ways, functioning as a repository of curriculum revisions, including curriculum research, measurement of effectiveness, best practice, lessons learned, and a set of pedagogy and assessment techniques, including outcomes tracking, and faculty research and development opportunities.

A commonly used tool in KM is a Knowledge Management System (KMS), which is an information technology (IT) based information system used to assist in the management of organizational knowledge. In this research, the definition of a KMS is that it is a tool to facilitate a knowledge flow and enables knowledge provision and sharing among knowledge suppliers (professionals and experts) in the form of Communities of Practices (CoP). CoP comprise a set of people who share a concern, a set of problems, or are passionate about a topic, and strive to deepen their knowledge and expertise in this area by interacting on an ongoing basis [8].

3.2.3 A Knowledge Supply Chain (KSC): Integrating SCM and KM

Although research has demonstrated success in applying KM tools to higher education, limitations are still found. Practitioners Thitithananon and Klaewthanong [9] claim that:

- It is difficult to classify and assess knowledge in curriculum design.
- The social network and CoP for curriculum are difficult to reach.
- There is difficulty in labeling knowledge as a product which can be moved.
- The time to revise and update curriculum takes at least 3–5 years.
- It is both complicated and time consuming to apply KM to the whole process of curriculum design.

Based on the features and elements of the supply chain and the integration of knowledge management, the author defines the knowledge supply chain as, *"the knowledge flow and management process in the cycle of e-tourism curriculum design, implementation, and maintenance from suppliers to end-users."*

3.3 Closing the Gap Between Knowledge Supply and Demand in e-Tourism Curriculum Design

To close the gap between knowledge supply and demand in e-tourism curriculum design, a knowledge engineering approach is used to identify e-tourism knowledge requirements and how to supply that knowledge in a KSC. Four main steps were followed to identify and solve the two fundamental problems associated with e-tourism curriculum design. These steps are: Data collection and structuring; Benchmarking; Case modeling; and Case simulation.

3.3.1 Data Collection and Structuring

In-depth interview was used to capture knowledge, which is one of the most extensively-used methods of data collection. The interviewees who are from both academia and industry in this research went through structured interviews involving face-to-face questioning, emails, and telephone dialogues. Six professionals from the departments of Human Resources, IT, and Marketing of tourism sectors in Thailand were interviewed face-to-face in order to determine knowledge demand from the industry, for those departments extensively dealing with the development personnel and know about industry requirements. Twelve academics of three e-tourism relevant sections (Tourism, IT, and Business sections) from both research-intensive university (e.g., Chiang Mai University) and teaching-intensive university (e.g., Payap University) were asked about the questions of what and how knowledge is supplied in the e-tourism curriculum which covers tourism, IT, and business area. Finally, 60 senior students from three Departments (tourism, IT, and Business sections) were met to identify the gap between knowledge demanded and supplied. These students are doing or have just finished their cooperative education/internship program. Once this knowledge was captured, there was a need to structure it using KE.

To structure the collected data with regard to e-tourism curriculum design, the Common Knowledge Analysis and Data Structuring (CommonKADS) tool was used to specify a knowledge model, including task, inference, and domain knowledge. Task, inference, and domain knowledge can be described as follows:

- Task knowledge: the knowledge required to complete tasks or achieve desired goals when problem-solving or decision-making.
- Inference knowledge: the control of knowledge abstracted from the task that describes the steps (or reasons for the steps) in a problem-solving task.
- Domain knowledge: the conceptualization of knowledge contributing to problem-solving in a particular domain. Normally, domain knowledge is derived from experts as related to their learning, working, and problem-solving.

A CommonKADS question template was used to elicit the effects of the questions in terms of the questioning process (adapted from [4]):

Question 1: What is the most important inference? → To generate assertion into a rule.
Question 2: What is the first concept coming to your mind? → To generate assertion into a rule.
Questions 3: What are the important factors? → To generate more rules.
Question 4: Do you have some alternatives? → To generate more rules.
Question 5: Can you tell me more about that? → To generate further dialogue if the expert dries up.

The second stage involves knowledge analysis, where the needs from industry and academia are analyzed and balanced according to the development of tourism at global, regional, and local perspectives, in addition to the requirements of higher education.

The third aspect is knowledge validation, which teaches back the acquired knowledge to experts based on the transcripts of the interview. The experts can interrupt the teach-back at any point if they disagree, or can make corrections and add on additional information or discussion. In this way, knowledge captured and analyzed is validated.

The last step involves modeling the knowledge into structured diagrams (Figs. 3.1, 3.2 and 3.3).

The data was collected at two universities within Chiang Mai, northern Thailand, in order to assess e-tourism curriculum provision in the GMS area, which needs to modify its e-tourism curriculum provision to reduce the leakage of tourism revenue. Once data had been collected, structuring took place using CommonKADS. To

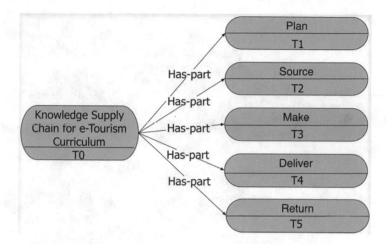

Fig. 3.1 Task knowledge modeling

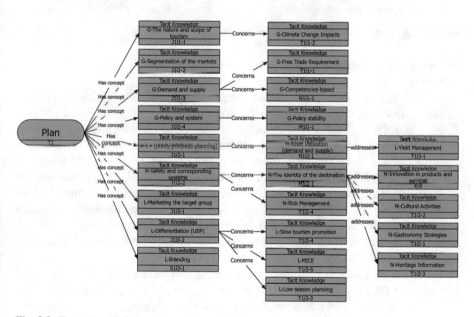

Fig. 3.2 Example of inference knowledge modeling

characterize the current knowledge flow from knowledge suppliers to users in curriculum design required consideration of both tacit and explicit knowledge. The domain knowledge from the experts was managed according to the stimuli-response in the interviews. Based on this, software requirements of the knowledge management system (KMS) were specified in preparation to model the domain knowledge

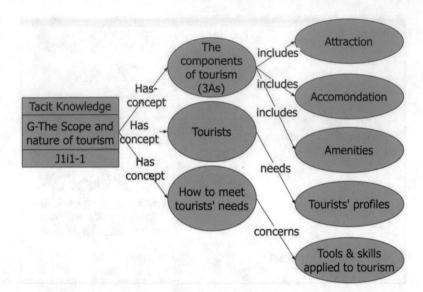

Fig. 3.3 Captured domain knowledge modeling

into a knowledge package to share among the three key actors of e-tourism curriculum design.

According to the supply chain operations reference (SCOR) model, the process of curriculum design was divided into five processes, which are Plan, Source, Make, Deliver, and Return, aiming to achieve the creation of a knowledge supply chain for e-tourism curriculum design. The templates of the CommonKADS used in this research provide faster and more effective method in task (Fig. 3.2) and inference knowledge (Fig. 3.3) modeling, which corporate tacit knowledge explicit as well as to make use of explicit knowledge. By using knowledge templates, it directly supports systematic and structured modeling, which generate a graphical representation from the knowledge captured in the interviews and documentations. Figure 3.1 shows the separation of the curriculum design task into these five SCOR processes using CommonKADS.

Based on the in-depth interviews with experts and professionals, the structured CommonKADS template was used to sort transcripts according to the priority of the inferences in the stimuli-responses. Taking "Plan" as an example, questions were asked in three groups, namely; why (e.g., the most important issues when designing an e-tourism curriculum from a global view, and *why* are they important?), what (e.g., *what* is the nature of the tourism of Thailand/Chiang Mai?), and how (e.g., *how* knowledge can be implemented in a computer system?).

Figure 3.2 illustrates for the example of "Plan," how questions were asked and how the resulting data was structured using a CommonKADS framework. The knowledge captured was coded and structured according to different knowledge providers (the interviewees), e.g. for the same question, the code J1i1 means the knowledge captured from a tourism professional J, which concerns the knowledge from a tourism curriculum designer T, coded as T1i1-2. In this way, it is easier to analyze and validate the knowledge from different experts for further knowledge modeling.

The captured domain knowledge further explained the steps and reasons contributing to solving the e-tourism curriculum design issues. Figure 3.3 shows an example of this domain knowledge modeling. Among all the inference knowledge captured and analyzed within the SCOR model's "Plan" step of e-tourism curriculum design, the scope and nature of tourism was the first factor considered by the professionals who were interviewed. This was then separated into the three components of tourism (3As: attraction, accommodation, and amenities; tourists; and how to meet tourists' needs). Therefore, when planning the curriculum, these inferences must be considered as a priority.

Structuring the data using CommonKADS identified three main problems with e-tourism curriculum design and how they might be solved.

- Problem one is a lack of knowledge provision and sharing and is believed to directly influence problems two and three.
- Problem two is the long lead-time for the process of curriculum design and implementation and directly results in the mismatch between society needs and curriculum provision.
- Problem three is that a disconnection lies between the curriculum designer and curriculum operator, which separates the planning of curriculum content from industrial requirements.

As the three problems identified above, the lack of knowledge provision and sharing is the most-often mentioned issues when designing an effective curriculum. Long lead-time is another important reason, which results in the gap between society needs and curriculum provision. The disconnection between curriculum designers and operators was identified by four senior lecturers/curriculum designers. Following the data collection and analysis, there was a need to benchmark the existing e-tourism curriculum in the case study against other e-tourism curricula, both inside and outside of the GMS area.

3.3.2   Benchmarking

Curriculum comparison and analysis was conducted by collecting best practice information regarding e-tourism curriculum from literature and the Internet and comparing among 11 universities in America, Europe, and Asia. These universities were selected based on their comparatively good reputation (determined by citations in research), which are: *Through data collection and benchmarking*, it was found that no universities in Thailand provide e-tourism curriculum as an independent discipline. Although much progress has been made in America (e.g., Temple University), Europe (e.g., Bournemouth University), and Asia (e.g., The Hong Kong Polytechnic University), the two problems identified in this research (the lack of knowledge provision and sharing and the long lead-time) have not yet been solved. Therefore, this research continued to utilize Chiang Mai University (CMU) as a case study to model the "as-is' situation based on current tourism curriculum design, and then simulate the desired "to-be" framework for e-tourism curriculum design at the College of Arts, Media and Technology, CMU.

3.3.3 Understanding the "As-Is" Situation

CMU, Thailand is one of the top three Thai universities and is a research-intensive university with the aim of becoming a comprehensive institution in northern Thailand for the sake of social and economic development of the region and the country as a whole. Structured interviews with a set of prepared questions in a form of knowledge template were conducted to conceptualize and identify problems. Many issues were found, but, this research focuses on the two major problems (illustrated in Fig. 3.1), which are:

- The long lead-time of the curriculum, which often takes 4 years to achieve a complete turnaround in design and delivery, with slight changes after 2 years.
- The implementation of curriculum is a one-way flow, which shows a clear disconnection between curriculum designers and curriculum operators, which make knowledge provision and sharing difficult.

The determination of the "as-is" situation in turn allows the simulation of the desired "to-be" situation and ultimately the design of a Knowledge Supply Chain (KSC) to close the gap.

3.3.4 Simulating the Desired "to-Be" Situation

The College of Arts, Media and Technology (CAMT) at CMU has been involved in e-tourism practices and KM research since 2003. Moreover, CAMT participates in the Erasmus Mundus Action 2-"A sustainable e-tourism project (2010–2014). In 2009, CAMT started an e-tourism elective program for undergraduates who majored in Modern Management and Information Technology (MMIT). With sound knowledge of business management, competencies of technology, and intense training in e-tourism, this batch of students, termed "e-tourism electives," is regarded as possessing the most comprehensive and complete body of knowledge about e-tourism in CMU.

A KMS was established in CAMT in 2004, and has been run as a platform for knowledge provision and sharing. This system was then developed specifically and utilized in e-tourism curriculum design and development in 2012 by creating Ten Communities of Practice (CoPs), which have been set up as a knowledge warehouse to share, store, retrieve, and communicate knowledge from knowledge suppliers to users. In CAMT, the KMS provides a mechanism to manage the tacit and explicit knowledge of 10 CoPs. These ten CoPs can be categorized into three main groups [10], including:

- A *community for knowledge creation*, which is a small community of experts or researchers (less than 20 people) assigned membership by pooling disparate knowledge to look for new ideas and new knowledge through discussion and dialogue;
- A *community for practice*, which consists of a core of experts who cover reasonably well-established knowledge with assigned roles to build assets, define standards, and seek best practice; and,
- A *community of capability stewardship*, which has a few subject matter experts, assigned to the community to maintain and update the standards by training, including coaching as well as monitoring how standards are applied.

During the structured interviews, the priority of knowledge in problem-solving indicated the inference and domain knowledge that experts employed as desired solutions in problem-solving. Knowledge structured and modeled through CommonKADS provided knowledge and filled up the gap between industry needs and curriculum provision. A KMS of e-tourism curriculum design functions as a platform of knowledge provision and sharing for 10 CoPs to contribute and communicate their domain knowledge. Through the integration of this KMS with the SCOR model, a knowledge supply chain of e-tourism curriculum design is proposed to solve the two fundamental e-tourism curriculum design problems, and is shown in Fig. 3.4.

As Fig. 3.4 shows, the model is organized around the five primary processes of *Plan, Source, Make, Deliver,* and *Return* which incorporate all interactions from knowledge supplier to knowledge user in the process of curriculum design, implementation, and development. From the KSC, knowledge suppliers such as curriculum designers, experts, and researchers, as well as professionals will be more aware of the holistic process of knowledge flow and the interactions between each of the five processes. The KMS implementation was supervised by knowledge managers who will help to facilitate the information flow to shorten the lead-time of the curriculum. In this way, the lack of knowledge provision and sharing can be solved by communication and activities of the CoPs within the KMS. The expected outcomes ("to-be" situation) of this article are summarized below:

- SCOR model captures and divides the process of curriculum design into five steps; and KMS connects lecturers, researchers, and professionals in a form of CoP and allocates these CoPs to the five steps.
- Knowledge is planned on the basis of customer knowledge through close communication and interaction among CoPs of lecturers, researchers, professionals as well as policy and quality audits.
- Knowledge is captured and recorded through CommonKADS method from contribution of CoPs on KMS functioning as a knowledge warehouse where updated knowledge can be provided, shared in a sustainable way.

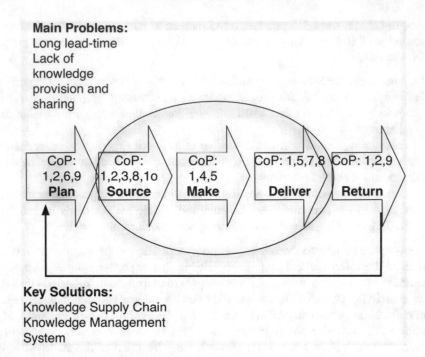

Fig. 3.4 A SCOR model for Knowledge Supply Chain (KSC) operated on a KMS

- With the communication, interaction and cooperation among CoPs of researchers, lecturers, and professionals on KMS, knowledge can be effectively made, delivered, and updated.
- Knowledge feedback will be brought to KMS and curriculum planners by CoPs of researchers, customers, and quality audits to shorten the lead-time of the update and renewal of the curriculum to meet both the academic and industrial requirements.

It is indicated that the innovations and synergy of KM tools with SCM can significantly enhance e-tourism curriculum design to reach the desired "to-be" situation.

3.4 Conclusions

The evolution from tourism to e-tourism requires higher education to adjust its intellectual outputs to match market and society needs. Curriculum, as the core factor in the quality of educational outputs, is therefore a key part of this transformation. Practices of knowledge management or supply chain management have been applied to enhance the efficiency and effectiveness of higher education, but still cannot fill the gap between higher education and society needs. This research has proposed knowledge engineering approach to develop a knowledge supply chain, defined as, "*the knowledge flow and management process in the cycle of design, implementation, and maintenance from suppliers to end-users.*" In running this knowledge supply chain, knowledge management and supply chain management are integrated for the first time. This knowledge supply chain is standardized via the use of the Supply Chain Operations Reference (SCOR) model into the five essential processes of plan, source, make, deliver, and return. The case study, Chiang Mai University, Thailand in the Greater Mekong Sub-region (GMS) was chosen and investigated.

To solve the two major problems in previous e-tourism curriculum research, namely, the long lead-time of the curriculum and the lack of knowledge provision and sharing, a knowledge engineering approach was applied as a powerful way to model and facilitate the knowledge flow in the knowledge supply chain of e-tourism curriculum. This research is the first step in using a knowledge supply chain in e-tourism curriculum development and also presents opportunities for further research. Such further research could assist GMS developing countries in matching the knowledge needs in the tourism industry, with that supplied by higher education and could ultimately increase tourism revenue by preventing tourism revenue leakage.

References

1. Mistilis, N., Buhalis, D., Gretzel, U.: Future e-destination marketing: perspective of an Australian tourism stakeholder network. J. Travel Res. **53**, 778–790 (2014). 1–13
2. Asian Development Bank Report: Greater Mekong Sub-region: Tourism Sector Strategy. (2012). Available at http://www.adb.org
3. Jenkins, C.L.: Tourism policy and planning for developing countries: some critical issues. Transp. Res. Record J. (2014)
4. Schreiber, G., et al.: Knowledge Engineering and Management: The CommonKADS Methodology. The MIT Press, Cambridge (2000)
5. Yen, M.: Customize GIS education with SCM model. In: Proceedings of ESRI International Conference (2005). Available at http://gis.esri.com
6. Petrides, L.A., McCleland, S.I., Nodine, T.R.: Costs and benefits of the workaround: inventive solution of costly alternative. Int. J. Educ. Manag. **18**(2), 100–108 (2004)
7. Kidwell, J.J., Vander Linde, K.M., Johnson, S.: Applying corporate knowledge management practices in higher education. Inf. Alchemy Art Sci. Knowl. Manag. **2**, 1–24 (2001)
8. Wenger, E., McDermott, R., Snyder, W.M.: Cultivating Communities of Practice. Harvard Business School Press, Boston (2002)
9. Thitithananon, P., Klaewthanong, T.: Knowledge management is a perfect education development tool. J. Knowl. Manag. Practice. **8**, 25–46 (2007)
10. Saint-Onge, H., Wallace, D.: Leveraging Communities of Practice for Strategic Advantage. Butterworth, Heinemann, Boson (2003)

Chapter 4
Improvement of Tourists' Experience to Promote Sustainable Tourism

4.1 Benefit of Inconvenience and Sustainable Tourism Design

In this section, a concept for designing systems called Benefit of inconvenience (BI) and a communication scheme called Media biotope (MB) are introduced in order to discuss sustainable tourism from an aspect of communication.

In the concept of BI, various benefits of inconvenient things/events are utilized for designing explicitly and actively. In the scheme of MB, awarenesses of surroundings through interactions with the environment are focused on. By using design criteria of BI and communication model of MB, realizing sustainable tourisms is expected.

4.1.1 Benefit of Inconvenience

Kawakami has proposed a concept for system design, called Benefit of inconvenience (BI)[1] [1].

Generally, "saving time and labors" is considered as an important goal when we are designing something, and many industrial products have been designed based on the criterion. However, we need another principle to design tourism services because services such as "saving time and labors" are not always satisfy tourists [2]. For example, people enjoy camping in natures, e.g. forests and lakesides, where do not have convenient facilities. Some people go so far as to visit backwoods where difficult to approach. Many tourists are also willing to join experience-based

---

[1] In the original paper, it is called as Fuben-eki (FUrther BENEfit of a Kind of Inconvenience). In this book, the author use Benefit of inconvenience (BI) for brevity.

© The Author(s) 2021
K. Soratana et al., *Supply Chain Management of Tourism Towards Sustainability*, SpringerBriefs in Environmental Science, https://doi.org/10.1007/978-3-030-58225-8_4

activities of creating local crafts in spite of the fact that they can purchase them in souvenir shops easily.

In order to discuss the situations, the concept of BI is well-served. The basic idea of BI is utilizing the following positive effects of inconveniences for designing systems.

- Fostering affirmative feelings.
- Providing motivation to tasks.
- Providing personalizations.
- Putting users at their ease.

In order to design BI based system (BI system), the following three principles and a guideline have been shown [3]:

Make visible:　It helps users understand systems.

Enhance awareness:　It allows users exploration.

Motivate creativity:　It allows users to do creative contributions.

Figure 4.1 illustrates the guideline. In the guideline, systems are arranged onto a two-axis space, convenience–inconvenience and benefits–harms. In this figure, arrows labeled (A), (B), (C), and (D) show basic ideas to realize BI system. Hasebe et al. have proposed a supporting tool for deigning BI system based on the guideline [4].

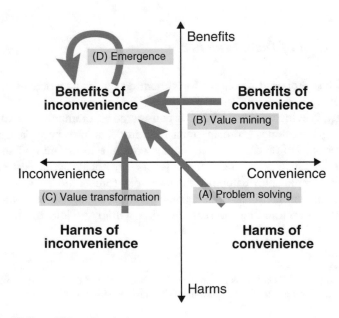

Fig. 4.1 A guideline of BI system design

4.1.2   Media Biotope

4.1.2.1   Basic Idea of Media Biotope

Developing the Internet communication technologies makes it possible for us to communicate with people in the distance easily and quickly, and our life has become "convenient." Meanwhile, these technologies weaken the connections between residents of local communities [5]. As a result, local residents often become indifferent to others, and activities in regional societies are declining.

As an approach to the problems, the concept of Media biotope (MB) has been proposed [6]. In the concept of MB, structures of communication media are discussed using an analogy of biotope in nature world (eco-biotope). Biotope stands for small areas which are suitable for living things, e.g. a marshland, a community-based forest, or a grassy meadow. These biotopes have unique ecosystem, but they are not isolated. They are relating to each other by the contributions of small living things, e.g. wild birds and insects, which travel between the biotopes.

MB can be considered as biological system of information media. Figure 4.2 illustrates the concept of MB. Upper part of Fig. 4.2a shows a uniformed mountain. In this environment, trees block sunlights, and undergrowths cannot survive. Then, the environments are uniformed. In Japan, a lot of Japanese cedars had been planted after World War II. As a result, many mountains lost the individualities and became like this. On the other hand, upper part of Fig. 4.2b shows biotopes which have diverse individualities. Many types of plants are growing and they create multi-layered ecosystem. In biotopes, undergrowths grown richly, and diverse living things which live there create sustainable ecosystems.

The lower part of Fig. 4.2 shows structures of communities formed by communication media. The left-hand side stands for a community which is formed by a

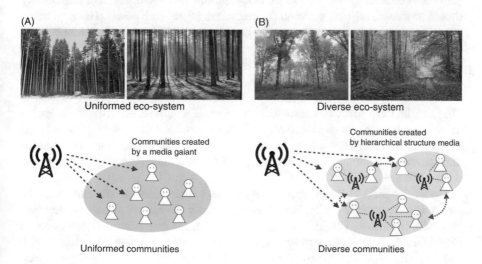

Fig. 4.2 Concept idea of Media biotope

media giant. As same as undiversified mountains, this type of communities tends not to have value diversity. The right-hand side stands for a structure of MB. In the structure, various types of communities are created by small local media, e.g., cable television, free papers, and community FM radio, and the communities are connected with inter-community media. Namely, different types of media make a multi-layered structure as same as biotope in ecosystem. In MB, each community keeps the originality and sustainability.

In this book, mediums that create MB are called as biotope-oriented medium and communication systems which include biotope-oriented mediums are called as MB system. For example, in the lower part of Fig. 4.2b, each symbol of antenna stands for a biotope-oriented medium, and the network consisting of several types of mediums is a MB system.

Obviously, biotope-oriented mediums are "inconvenient" in comparison with media giants because they cannot transmit informations for large area quickly. However, the inconvenience connects local people and creates a sustainable community. In this meaning, biotope-oriented mediums can be considered as implementation of BI.

4.1.2.2 Characters of Media Biotope

Suto defined Media biotope based on a definition of eco-biotope [7]. First, Media biotope has the following five characters:

Smallness: Here, "smallness" has two meanings; physical scale and mental scale. First, physical proximity of members of the communities must be comparatively small. In a media biotope, members of a community should be possible to have face-to-face communication together too. In other words, communication media help real communications in media biotopes. Second, members of a community should be easy to utilize the media for communications with others. Here, "utilize" means not only obtaining informations but also working on the other members by using the media. In addition, the influence of each person should be ensured. It brings the members to continuous activities, and sustainable communities are created.

Connectivity: Communities must be connected loosely by upper layer media. Thus, media must form a hierarchical structure. If a community created through a medium with a scale as described above becomes isolated from other communities, it might be dominated by a specific sense of values. This means that communities must be able to communicate each other so that communities can respect the others' sense of values.

Openness: Media biotope must have openness such as any people can affect the media and community easily. For instance, usually we cannot affect on television programs broadcasted from key stations easily. Such television programs can be thought as an analogy of controlled-access nature reserved area in the nature world, and they cannot be considered as media biotopes. Gated communities [8] also cannot be called a media biotope because it is separated from the environment as same as well-kept home gardens. In media biotopes, communities must have interactions with the environments surrounding them and keep relationships with them.

Intention: Media biotope must be designed by residents based on a concept. Nowadays, we have a lot of convenience communication ways, and there is a tendency for accepting the latest technology media. Sometime, we even feel a kind of obsession that we have to use these latest media. Thus, we have to discuss that what is the suitable communication medium for the region to create a sustainable community, and have to try to realize the medium.

Autonomia: Communities as a media biotope must be autonomous system as with eco-biotopes. The autonomous systems should not be kept by effects from outside the systems but by the results of the systems' behaviors. That is, informations should be circulated in the systems instead of being simply broadcast and consumed. Here, communities should be able to adjust in accordance with changes outside the communities while maintaining their autonomy.

This character is also an important factor for designing sustainable tourism systems.

4.1.2.3 Communication Model for Media Biotope

Two schemes of communication through a medium are illustrated in Fig. 4.3.

(A) Shannon and Weaver's model for communication media.

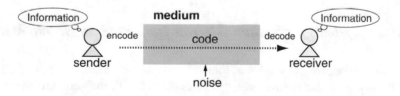

(B) Author's model for communication media.

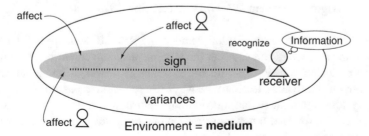

Fig. 4.3 Models for illustrating communication mechanisms

Figure 4.3a shows the traditional scheme of communications which is shown by Shannon and Weaver [9]. The gray square between persons stands for a communication medium, and the arrow with dashed line means an information flow. In this scheme, an information sender and a receiver are included. First, an information held by the sender is encoded and transmitted to the receiver through the medium. The information is then decoded, enabling the receiver to understand it. In this scheme, accurate reproduction of the original information is the main concern

As we can see in the figure, situations in which communication channel has connected between a sender and a receiver can be represented by using the model. However, tourists do not necessarily establish communication channels around them. In many situations, they do not obtain informations positively. They just feel the sceneries surrounding them and enjoy them. Thus, we need another model for representing communications of tourists.

Figure 4.3b shows a new model for communications proposed by Suto et al. [10]. In this model, the gray oval means a communication medium, white oval means the ambient environment of the receiver, and the arrow with dashed line means a flow of information. First, signs included in the environment are transmitted through the medium to the receiver. Then, the receiver obtains informations by recognizing signs on the basis of his/her context. That is, the receiver understands the ambient environment of him/her with signs obtained through the medium. In this scheme, information senders are not included. Consequently, above described situation in which a tourist enjoys scenery around him/her can be illustrated with this model

The new model has backward compatibility for the traditional model, and it can illustrate situations which are represented by the traditional model.

4.1.3   Systems for Sustainable Tourism Based on BI and MB

In this section, four examples of tourism systems which can be considered as BI/MB systems. Their mechanisms can be explained with the communication model described in the previous section.

4.1.3.1   Navigation System Which Does Not Show Details

Usually, navigation systems display detailed maps to show users the ways to the destinations clearly. We can say that the systems are convenient because the users can get the destinations easily with the systems. However, it has a side effect of atrophying our motivation to recall the ways where we have passed. As a result, the users pay less attention to the surrounding environment. To solve the problem, Nakatani et al. proposed a new concept navigation system for tourists who enjoy walking in sightseeing areas [11].

As mentioned above, usual navigation systems are convenient for persons who have already had explicit destinations. Meanwhile, Nakatani et al. focused on roaming style tourisms. In this style of tourism, tourists walk in the areas while enjoying the environment.

Figure 4.4 illustrates basic operating procedures of the navigation system.

(A) Step 1

Some sightseeing spots are displayed on the screen.

(B) Step 2

Draw a path on the map by referring the informations displyed on the screen.

(C) Step 3

Details of the map disapper after starting the navigation.

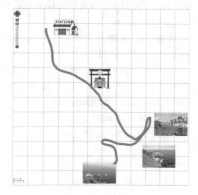

Fig. 4.4 Screens of the navigation system which does not show details

1. A map showing the area around the user is displayed on the screen with several photographs of sightseeing spots (Fig. 4.4a). Users can make a plan of course by watching the screen.
2. Users draw a line from the departure place to a destination on the screen by using a finger (Fig. 4.4b). The line represents the course which the user wants to enjoy walking.
3. When navigation starts, the detailed map disappears and only the hand drawing line is shown on the screen (Fig. 4.4c). The user visits the sightseeing spots by referring only to the line on the screen.

Nakatani has defined the word "tourism" as "leaving from the daily living area and interacting with the visiting area in vacation time." In this definition, "interaction with environment" plays an important role. From the aspect of convenience, the novel navigation system could have problems. However, it can encourage users to interact with surrounding environment. As a result, users get many signs from the environment, which are recognized as information.

If they do not pay attention to the environment, they are easy to lose the way. At first glance, it seems inconvenient. However, it increases chances to find something which is not on the way they planned in advance. Thus, the system can be considered as an example of (A) problem-solving in Fig. 4.1.

4.1.3.2 Degrade Navigation System

Degrade navigation system is another solution proposed by Kitagawa et al. [12]. First, the users can use the navigation system in the same manner as normal navigation systems. The different point is that the system erases the map gradually along the routes which the users have passed. Eventually, the routes where the user used repeatedly vanish. Thus, the system can be called as "Navigation system with blurring map."

Figure 4.5 shows examples of screens of the system. The left map shows a region where the user's office is located. The office room is located in the area which is mostly erased. The right is a small-scale map of the same region. Some paths are gradationally erased in the map.

The system encourages users to observe the surrounding environment more carefully and to recall landmarks in the environment more precisely. By using this system, the sceneries became more memorable ones for tourists. In this meaning, this system seems similar with the first example. However, the designing concept is different. In this case, functions of personalization and encouraging observation are given for a usual navigation system. Thus, the system can be considered as an example of (B) value mining in Fig. 4.1.

Fig. 4.5 Sample screens of degrade navigation system

Fig. 4.6 Images of community activities on a bus

4.1.3.3   Communicative Buses

Communicative bus is a conceptual idea of commuting system in sightseeing spots [7]. Figure 4.6 shows images of community activities on a "communicative bus." The system has three main functions; commuter for residents, transportation for travelers, and place for communication between residents and travelers.

Several local products, e.g., lunchboxes, sweets, and souvenirs, are sold to the travelers at each stop (Fig. 4.6a). Passengers can enjoy lunch or a snack on the bus (Fig. 4.6b). The name of the next stop and products sold there are announced in the bus before each stop. Tourists would get a chance to communicate with the local people by buying their products. In this way, the tourists would naturally improve their understanding of the region while enjoying their trip.

Electronic message displays could be set up on the front of the buses, on the side near the back doorway, and above the driver (Fig. 4.6c). Passengers can send brief messages like Twitter messages about their impressions, memories, etc., of the trip and about local products to these displays by using their smartphones. By posting messages, passengers can share their thoughts and feelings with people outside the bus.

Consequently, their trip would become more impressive. Usually, it takes long to get somewhere by using local buses, and it is considered as inconvenience. However, tourists can enjoy a relaxing trip with local buses owing to this system.

This system can be considered as an example of (C) value transformation in Fig. 4.1 because the inconveniences of "slow" of local commuters are utilized for obtaining novel values.

4.1.3.4   Roulette Tour

A new style tourism, "Roulette tour," is getting popular in Kyoto, Japan. In the center of Kyoto city, main streets follow a grid pattern. The tour is a kind of game which uses the unique structure of the city.

The participants of the tour make parties. Each party is given a roulette on which four directions, north, south, west, and east, are printed, and carries it with them. Each Party starts from a predetermined point in Kyoto city. When they meet an intersection, a member of the party spins the roulette wheel to decide which direction they are going to. That is to say, the participants cannot choose a direction of travel by themselves.

Sometime, they must go back and forth on a street by following the results of roulette. Although it seems quite inconvenient for sightseeings, a lot of tourists are willing to join the tour. Why does the tour attract many tourists? To consider the reason, we have to think "the purposes of tourism."

Sometimes, the purpose of travels tends to be considered as "visit" to the destinations. In this sense, once a tourist has visited a place, the value of the place for the tourist decreases because he/she has already achieved the goal. It can be considered as a resource-consumptive tourism. On the other hand, "process" and "experiences" are focused on in the concept of BI. Various experiences in the processes make the travel very unique and memorable. It can be considered as sustainable tourism.

In the case of the Roulette tour, "destination of the tour" has been discarded, and experiences on the processes are highlighted. As a result, the participants can have unique experiences and they think that they want to join the tour repeatedly. The idea has been emerged as an analogy of Japanese game of "sugoroku" and the grid-like structure of Kyoto city. It can be considered as an example of (D) Emergence in Fig. 4.1.

4.1.4   Conclusions

In this chapter, two concepts, i.e. Benefit of inconvenience (BI) and Media biotope (MB) have been introduced for discussing sustainable tourism system from an aspect of communications. Then, four ideas for tourism have been shown as examples of BI/MB based systems. These systems work as inter-media which connect local people and tourists as shown in Fig. 4.7. In this figure, the orange-colored dome stands for a local community, and blue-colored domes stand for communities of tourists. The yellow-colored dome stands for an inter-medium which connects among the communities. As a result, these communities form a hierarchical structure. Thus, the inter-media can be considered as biotope-oriented mediums. In this way, sustainable media biotopes are built.

The author believe that these concepts can be hints for designing new sustainable tourism system.

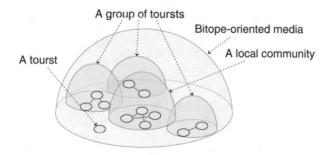

Fig. 4.7 Media biotope created by BI and MB systems

References

1. Kawakami, H.: Further benefit of a kind of inconvenience for social information systems. Lecture Notes in Computer Science, pp. 301–306. (2013)
2. Watanabe, W.C., Patitad, P., Suto, H.: Tourists' satisfaction: benefits of inconvenience aspect. In: Proceedings of the Fourth International Conference on Electronics and Software Science, pp. 134–143 (2018)
3. Kawakami, H., Hiraoka, T.: Methods for designing systems with benefits of inconvenience. In: HCII 2019, LNCS 11583, pp. 253–263. Springer Nature, Cham (2019)
4. Hasebe, Y., Kawakami, H., Hiraoka, T., Nozaki, K.: Guidelines of system design for embodying benefits of inconvenience. SICE J. Control Meas. Sys. Integr. **8**(1), 2–6 (2013)
5. Dreyfus, L.H.: On the Internet. Routledge, Abingdon (2001)
6. Mizukoshi, S.: Media Biotope (Japanese). Kinokuniya Publication, New York (2005)
7. Suto, H.: Media biotope: media designing analogous with biotope. In: 2010 International Conference on Computer Information Systems and Industrial Management Applications (CISIM), pp. 75–80 (2010). https://doi.org/10.1109/CISIM.2010.5643688
8. Blakely, E.J., Snyder, M.G.: Fortress America: Gated Communities in the United States. Brookings Institution Press, Washington, D.C. (1999)
9. Shannon, C.E., Weaver, W.: The Mathematical Theory of Communication. University of Illinois Press, Champaign (1949)
10. Suto, H., Taniguchi, T., Kawakami, H.: A study of communication scheme for media biotope. In: SICE Annual Conference 2011, pp. 206–209 (2011)
11. Nakatani, Y., Tanaka, K., Ichikawa, K.: New approach to a tourist navigation system that promotes interaction with environment. In: Schmidt, M. (ed.) Advances in Computer Science and Engineering, Chap. 18. IntechOpen, Rijeka (2011). https://doi.org/10.5772/15661
12. Kitagawa, H., Kawakami, H., Katai, O.: Degrading navigation system as an explanatory example of "benefits of inconvenience". In: SICE Annual Conference 2010, pp. 1738–1742 (2010)

Chapter 5
Life Cycle Assessment and Supply Chain Management for Tourism Enterprises

5.1 Combination of LCA and SCM Aspects for Sustainable Development

LCA and SCM have their own strengths and limitations. LCA requires comprehensive inventory data in identifying hotspot, hidden cost, and tradeoffs. LCA considers environmental impacts contributed over the life cycle of a product or service. The life cycle of a product or service consists of four stages, which are raw material extraction, manufacturing, use, and disposal. SCM requires significant level of stakeholder involvement in increasing an organization's effectiveness. SCM helps reduce an organization's cost and support customer's demand. Details on how SCM and LCA can be incorporated in tourism industry are discussed in Chaps. 1 and 2.

The combination of LCA and SCM can uphold each other strengths and lessen each other limitations. Both LCA and SCM can be adopted by an organization to reduce cost via different mechanisms. LCA reduces cost of a product or service by identifying environmental hotspots which are generally caused by resource consumption. SCM reduces operation's cost by improving organizational activities. One of the common tools used in SCM is SCOR model, which was developed by the Supply Chain Council. Six categories of activities considered in SCOR model are plan, source, make, deliver, return, and enable. LCA can be applied to all six activities in SCOR model. During the planning, LCA can be used to aid in decision-making process. For other categories, LCA can be used in improving process or operation to reduce resource consumption and cost.

The combination of LCA and SCM is an appropriate approach to drive tourism towards sustainability. The tourism industry needs not only to serve the needs of visitors and local community, but also to be economically viable and environmentally friendly. Different kinds of tourism enterprises can operate their businesses following the LCA-SCM framework. Resources and/or operations used in different sectors across the tourism industry or among suppliers can be shared to reduce the

© The Author(s) 2021
K. Soratana et al., *Supply Chain Management of Tourism Towards Sustainability*, SpringerBriefs in Environmental Science,
https://doi.org/10.1007/978-3-030-58225-8_5

costs of resources and operation. Also, environmental impacts from products/services and logistics activities can be reduced. For example, in creating a local or seasonal menu of a restaurant to serve tourists, green sourcing (SCM) can be used along with LCA. LCA can be used to quantify environmental impacts of purchasing raw materials from different sources. Raw materials from sources with the most cost effective and minimum environmental impacts are chosen. Green logistics (SCM) option should be selected to transport raw materials. LCA can be applied to investigate hotspots of the option. Then, the transportation can be improved upon the LCA results. According to the LCA and SCM concepts, the new LCA-SCM framework of tourism is proposed. The framework is depicted in Fig. 5.1.

Referring to the proposed LCA-SCM framework for tourism, processes in the tourism sector in second Tier, except waste recycling disposal, are considered as inputs or raw material acquisition phase of LCA. Waste recycling disposal is

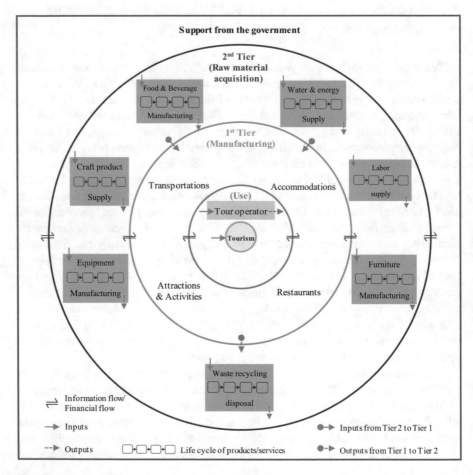

Fig. 5.1 Proposed life cycle assessment-supply chain management framework of the tourism industry

considered as a system to deal with outputs or end-of-life phase of LCA. Each process also has its inputs and outputs. For example, inputs of the food and beverage manufacturing process are mainly agricultural products with, e.g., organic wastes and leachate as outputs. Inputs of equipment manufacturing are mainly metals and electronic components, with, e.g., scrap metals and chemical wastes as outputs. The first Tier of tourism supply chain consists of four sub-sectors, which are transportation, accommodation, attractions and activities, and restaurants. The four processes are considered as manufacturing phase in LCA. Tour operator is considered as use phase in LCA, while tourism is considered as the main output, or product of the system. There are financial and information flows across the tiers. Also, every process should be supported by the government policies in driving tourism towards sustainability.

5.2   The Opportunities and Limitations of LCA-SCM for Tourism Enterprises

Using LCA or SCM alone may lead to environmentally friendly industry or economically viable industry, respectively; whereas, using the combination of both LCA and SCM would lead to sustainability. Tourism enterprises should be operated in a way to achieve environmentally friendly, economically viability, and social responsibility. In this section, how LCA-SCM framework can be applied to each tourism enterprises, such as accommodation, restaurants, and transportation are presented.

5.2.1   Accommodation

Accommodation comprises several activities from, e.g., a hotel room, spa, pool, and restaurant. In this section, the supply chain of a hotel room is discussed. The products and services provided for a hotel room are, e.g., bedding, cleaning, toiletries, and snacks and beverages. From SCM perspective, completing a service for a sustainable hotel room requires sourcing, laboring, and transportation. For sourcing, sustainable and local products should be a priority. It will not only generate environmentally friendly hotel room, but also support community and reduce emissions from transportation of goods. After developing supply chain scenarios of a hotel room, LCA should be applied to aid in decision-making process in choosing the most appropriate sources of products and services. In brief, SCM is used to design possible scenarios using green sourcing and green transportation from various suppliers, while LCA is used to decide which scenario is the most appropriate one to drive the hotel towards sustainability. Assuming each activity in raw material acquisition stage (Tier 2) has three suppliers available, which would create three scenarios for each activity. Scenarios can be developed as presented in Fig. 5.2.

A hotel has power to drive the supply chain towards sustainability. The hotel as a consumer can adopt sustainable sourcing and procurement. Decision can be made based on comparative LCA results. Using LCA will avoid the hotel business to choose a product or service based solely on the manufacturing phase, but also the end-of-life phase of the product. For example, a glass water bottle (an input from Tier 2 to Tier 1) is provided rather than a plastic water bottle in order to eliminate the waste disposal cost since the glass bottle can be returned to the manufacturer. Referring to Fig. 5.3, the output flows from Tier 1 to Tier 2 are wastewater and solid wastes to be disposed of at wastewater treatment/solid wastes disposal facilities. Used glass water bottles are reversed to beverage manufacturer. The total environmental footprint is improved. Landfills, the main player at the end-of-life phase, generally operated by public sector, are benefited from the operation, since the quantity of solid wastes is lessened. The benefits should be allocated among the players involved in the operation. One of the benefit allocation methods is an

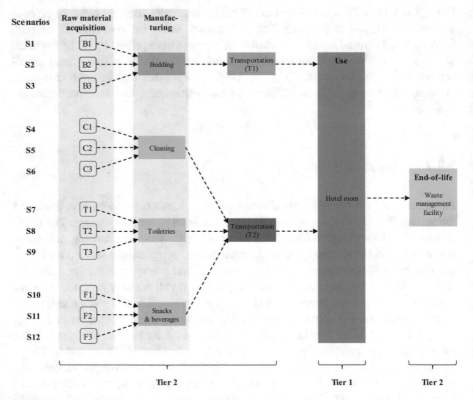

Fig. 5.2 Proposed life cycle assessment-supply chain management framework up to use phase of a hotel room

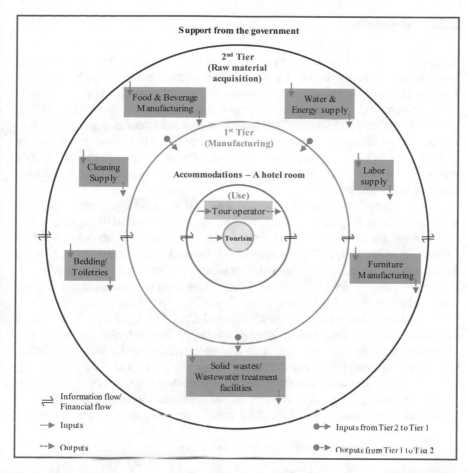

Fig. 5.3 Proposed life cycle assessment-supply chain management framework of an accommodation—a hotel room

incentive provided by the government to the hotel and the beverage manufacturer because they help reduce the operation cost and environmental impacts from landfilling.

Similar approach can be applied to a cleaning process of a hotel room. Cleaning a hotel room would require various types of detergents and chemicals. Local sourcing for chemical-free toilet cleaners, for bleaching, sanitizing, and scrubbing, can avoid cost for stocking and transporting. The hotel can build a network with local businesses to purchase locally-sourced products, which will also help boost local economy. In addition, environmental impacts from transportation of products are lessened. A group of hotels in the same area has power to drive their suppliers (local businesses) towards sustainability by implementing green sourcing. In this case, all players are in a symbiotic relationship business. The hotel and the local businesses can work together to serve each other's needs.

5.2.2 *Restaurant*

Using the LCA-SCM framework can drive the whole supply chain of a restaurant towards sustainability. The framework of sustainable restaurant SCM considers food material supply (e.g., producer, processor, and transporter) as upstream processes, purchasing, designing, storage, cooking, and service/packaging/marketing as middle stream processes, and cleaning and post-treatment as downstream processes. From a life cycle perspective, the system boundary in evaluating a restaurant from farm to fork includes processes such as food procurement, food storage, food preparation and cooking, food service and operational support, and end-of-life stage, as illustrated in Fig. 5.4.

From a life cycle perspective, inputs and outputs of each stage are varied. Inputs are, e.g., agricultural products, water, energy, and other supplies; while outputs are, e.g., wastewater, solid wastes, and air pollution. LCA practitioners have to collect inventories of each stage. The inventories will be used in LCI step of a comparative LCA, then to calculate for cost and environmental tradeoffs in LCIA step. After the most appropriate options are selected, SCM can be used to promote social involvement and fair income distribution.

Example of a restaurant case applicable to LCA-SCM framework is the use of compostable take-out packaging. The use causes positive and negative impacts on stakeholders in the supply chain. The packaging manufacturer will have higher sales. The restaurant that employs eco packaging will gain better corporate's image with additional cost, since the eco packaging normally costs higher than plastics or Styrofoam packaging. Environmentally conscious customers might have to pay for additional charge but might be satisfied with using eco-products. The landfilling

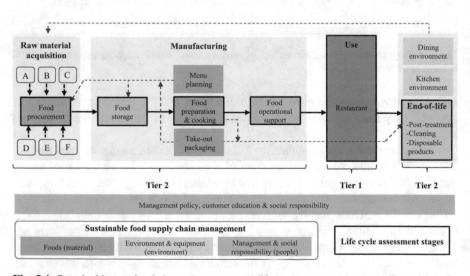

Fig. 5.4 Sustainable supply chain management and life cycle assessment system boundary of a restaurant (Modified from [1, 2]). The red line represents the relationship of different stakeholders in the food supply chain

facility will save cost spent on land use. In spite of the mentioned pros and cons of each stakeholder, everyone will gain advantage from better environment. Therefore, the pros and cons should be shared throughout the supply chain from the life cycle perspective to promote the green practices. In this case, LCA can be applied to quantify positive and negative impacts, either on environmental impacts or cost, while SCM can be applied to manage the operation (e.g. sourcing and logistics) among the stakeholders.

5.2.3   Transportation

Transportation in the tourism sector contributes the highest CO_2 emissions compared to accommodation and other tourism activities [3, 4]. The total CO_2 emissions from air, road, and other transports contribute approximately 75% of the total CO_2 emissions from the tourism sector [5]. SCM of transport, which provides, e.g., affordable air travel, more convenient connectivity, new business models, and better visa facilitation, is the major driving tool to promote international and domestic tourism. However, the advantages arise with congestion and environmental adverse impacts, mainly due to the energy use and emissions. Air pollution and noise from congestion causes human health impacts and possibly deteriorates tourist attractions. Several attractions such as natural parks and cultural heritage sites prohibit the use of personal petroleum fuel-engine vehicles on sites.

In addition to the tailpipe emissions, there are other sources of emissions over the life cycle of transportation. The emissions also come from, e.g., vehicle production, infrastructure construction, petroleum refinery, or electricity generation and its supply chains [6], as illustrated in Fig. 5.5. Therefore, the tourism transportation sector should be engaged in the CO_2 emissions reduction process.

The red box defines a life cycle of an automobile from raw material extraction to manufacturing and use phases.

Transportation service providers in tourism sector can use the proposed framework when making decision on transportation options. This framework can help avoiding choosing the transportation option based solely on the use phase. Also, material use, manufacturing processes, fuel efficiency, and maintenance should be taken into consideration. Based on the proposed LCA-SCM framework for tourism (Fig. 5.1), raw material extraction, vehicle and engine manufacturing processes, energy production and infrastructure construction and operation in Fig. 5.5 are Tier 2 activities; while the use of transportation service is considered as Tier 1 activity of the transportation supply chain. Generally, end-of-life phase of an automobile is not included in the system boundary due to its complexity [6]. However, it is crucial to also consider the end-of-life phase of an automobile. One example on the importance of taking the end-of-life phase of an automobile into consideration is an electric vehicle. An electric vehicle has been advocated as a "zero-emission vehicle." The term is true if only use phase of the vehicle is considered. There are emissions generated during raw material extraction, manufacturing, and end-of-life

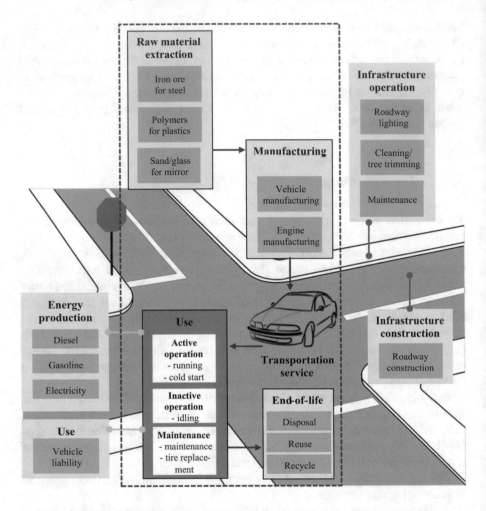

Fig. 5.5 A life cycle of an automobile and its supply chains

phases of an electric vehicle. Obsoleted batteries at the end-of-life phase of electric vehicles, which involve acid solutions and heavy metals, are needed to be properly managed.

There are other factors to be considered when making a decision on transportation option in the tourism sector. Such factors are types, travel distance, lifespan, load carrying capacity, passenger occupancy rate and maintenance cost of the vehicle, accessibility, and types of tourist destination. Each tourist destination can provide more than one transportation option, depending on, e.g., the geography and the number of tourists of the site. National parks encourage visitors to walk or bike on the route or to take a shuttle bus provided. The two options of transportation should be coupled with an effective map and appropriate time schedule with online reservation, respectively, in order to meet the need of visitors and to protect the destinations. Another fragile tourist destination is cultural heritage sites. Air pollution from transportation on site can potentially damage antiques and archeological sites. Historic Town Sukhothai in Thailand is one of United Nations of United Nations Educational, Scientific and Cultural Organization (UNESCO) World Heritage Sites that prohibits personal transportation entering the site. The historic site offers electric vehicles, both electric shuttles operated as schedule and electric cart for personal rent. Both options do not post negative impacts on site, however they consume electricity. Emissions from electricity generation negatively affect the community in the neighborhood areas of the power plant. The options might be appropriate in this case, Historic Town Sukhothai, since Thailand relies mainly on hydropower, which generates less greenhouse gas emissions comparing to other sources of electricity. However, LCA should be conducted to quantitatively support the decision.

LCA in combination with SCM could tremendously promote a more sustainable transportation service in the tourism sector. Efficiency and sustainability of transportation used to be sufficiently improved through adoption of technology innovations, such as hybrid system and alternative and renewable fuels [7]. Currently, other tools to promote sustainability in tourism transportation are business-to-business (B2B) relationship collaboration and a car-sharing business model. An example on a B2B collaboration is a sharing of airport shuttle buses among hotels in the same area. The approach incorporates SCM to make profit and provide services to customers. For a car-sharing business, the term can be defined as a short-term rental of a car or sharing of a car for a given trip. The former definition is referred to in this section. The business model shifts customer's demand from purchasing a car to purchasing car-sharing service. With the car-sharing system, the cost of purchasing and maintenance are shared among the business owner and the customers. In addition, the customers do not require a large amount of money to invest on a sparingly used car. The business model based solely on SCM serves the business financial concerns and customer needs. By incorporating LCA, car-sharing service can improve its cost and environmental performances, also. More fuel-efficient and cost-effective vehicles, from a life cycle perspective, are selected as rental cars. The LCA (environmental performance) of each car can be used to inform the customers as an environmental product declaration (EPD). In addition, the business plan can be improved based on a comparative LCA results on different car makes. For instance, tourism transportation enterprises can use LCA results in planning on which car makes to be purchased, or planning on when a rental car should be dismissed and sold.

References

1. Baldwin, C., Wilberforce, N., Kapur, A.: Restaurant and food service life cycle assessment and development of a sustainability standard. Int. J. Life Cycle Assess. **16**(1), 40–49 (2011)
2. Wang, Y.-F., et al.: Developing green management standards for restaurants: an application of green supply chain management. Int. J. Hosp. Manag. **34**, 263–273 (2013)
3. Michailidou, A.V., et al.: Life cycle thinking used for assessing the environmental impacts of tourism activity for a Greek tourism destination. J. Clean. Prod. **111**, 499–510 (2016)

4. World Tourism Organization and International Transport Forum: Transport-Related CO2 Emissions of the Tourism Sector – Modelling ResultsUNWTO, Madrid (2019)
5. World Tourism Organization and United Nations Environment Programme: Climate Change and Tourism – Responding to Global ChallengesUNWTO, Madrid (2008)
6. Chester, M.V., Horvath, A.: Environmental assessment of passenger transportation should include infrastructure and supply chains. Environ. Res. Lett. **4**(2), 024008 (2009)
7. Sarasini, S., Langeland, O.: Business model innovation for car sharing and sustainable urban mobility. Nordic Energy Research. (2017)

Printed in the United States
By Bookmasters